MW01627239

Documentary Media
Seattle, Washington

A VIEW OF THE METHOW

FROM MOCCASIN LAKE RANCH

James C. Pigott

A View of the Methow: From Moccasin Lake Ranch

Copyright © 2005 James C. Pigott
All rights reserved. No portion of this book may be reproduced or utilized in any form, or by any electric, mechanical, or other means, without the prior written permission of the publisher.

Published by Documentary Media
3250 41st Ave SW
Seattle, WA 98116
206 935-9292
books@docbooks.com
www.documentarymedia.com

First edition 2005

Printed in China

Written by James C. Pigott
Edited by Don Graydon
Publisher: Barry Provorse, Documentary Media
Book and Cover Design: Paul Langland Design, Poulsbo, Washington

Library of Congress Cataloging-in-Publication Data
Pigott, James C.
A view of the Methow : from Moccasin Lake Ranch / James C. Pigott.-- 1st ed.
p. cm.
Includes bibliographical references and index.
ISBN 0-9719084-9-4
1. Pigott, James C.--Family. 2. Ranchers--Washington (State)--Methow Valley--Biography. 3. Methow Valley (Wash.)--History. 4. Moccasin Lake Ranch (Wash.)--History. 5. Moccasin Lake Ranch (Wash.)--Biography. I. Title.

F897.M44P54 2005
979.7'28--dc22

2005002094

Photos on page 1 and 2-3 by Scott Chambers

Contents

Preface	**The Flow of Time**	11
Chapter 1	**Ancient Eden**	15
Chapter 2	**When Worlds Collide**	21
Chapter 3	**The Lure of Gold**	27
Chapter 4	**The First Settlers**	35
Chapter 5	**Unbreakable**	43
Chapter 6	**Rhythms of Life**	53
Chapter 7	**Taking Care of Business**	63
Chapter 8	**Growing Up**	73
Chapter 9	**A Ranch Is Born**	83
Chapter 10	**Thirty-Six Years of Ranching**	93
Chapter 11	**Hints of Things to Come**	99
Chapter 12	**A New Vision**	105
Chapter 13	**The Highway That Changed It All**	125
Chapter 14	**Ranch at a Crossroads**	131
Chapter 15	**Paths to the Future**	141
Chapter 16	**A Delicate Balance**	153
	Source Notes	160
	About the Author	161
	Index	162

Dedication

This is dedicated to the memory of my
father-in-law, Jonathan R. Titcomb.
His love of the wilderness led him to
purchase Moccasin Lake Ranch and
create a sanctuary where his descendants
and their families could experience
the outdoor life that he cherished.
This book would not have been written
were it not for his vision and foresight.

Shafer Museum, Winthrop, Washington, and the front porch of the Guy Waring cabin that locals still call The Castle. Photo by Richard Hart

Acknowledgments

I am indebted to those whose earlier research and writing have eased my task as an author. Much of the material in this book is drawn from newspaper articles, pamphlets, and books that provided historical accounts of life in the Methow Valley of northern Washington State. Dale Dibble's book *Methow Valley Pioneers* and articles from early editions of the *Methow Valley News* were especially useful. The Shafer Historical Museum in Winthrop assisted by providing access to photographs, personal articles, and machinery that helped tell the Methow story. My interviews with longtime Methow residents, descendants of the pioneers, and local historians were invaluable. I deeply appreciate their generous offerings of time and historical knowledge.

Aspen, bunchgrass, ponderosa pine, and in the distance fields of alfalfa. Photo by Scott Chambers

The Flow of Time

This book is a narrative of the Methow Valley and its people, from the time of the first human inhabitants more than seven thousand years ago to the early twenty-first century. There is a special focus on a representative example of valley history — the story of Moccasin Lake Ranch, which came into being a century ago and has been in our family since 1961, when it was purchased by Jonathan R. Titcomb, my father-in-law. The valley of the Methow (MEH-tauw), in north-central Washington State on the eastern flank of the North Cascade mountain range, is widely known today for its natural beauty and limitless recreational choices. It has become popular with outdoor enthusiasts as well as tourists eager to experience the spectacular scenery of the North Cascades and the Methow itself.

Neither my wife, Gaye, nor I had ever heard of the Methow Valley or Moccasin Lake Ranch until the day in 1961 that a letter arrived from Jon Titcomb. I was duty officer on board the Navy destroyer USS *Rupertus*, moored at the naval shipyard in Yokosuka, Japan. Midway through my watch on that quiet January afternoon, mail call sounded and I was given a letter addressed to me, Lt. j.g. James Pigott, and my wife, from Gaye's father. His letter carried the news that he was seriously interested in buying Moccasin Lake Ranch.

Gaye and I knew that Jon had been thinking of buying a ranch. He wanted a challenging project after retirement from the Weyerhaeuser Company — one that would address his love of the outdoors and rural life. Jon's attachment to the outdoors extended from his boyhood on a farm in Vermont, through his years at Dartmouth College in New Hampshire, to his corporate life at the forest products company.

As a boy, Jon often accompanied his father, a nationally recognized conservationist and New York State's first commissioner of fish and game, on trips into the wilderness. With encouragement from his father, he became an avid hunter and fisherman. At Dartmouth he was active in the Outing Club and in his sophomore year was named to the prestigious position of hutmaster at the college's famed Mount Moosilauke campsite. After earning a liberal arts degree at Dartmouth, Titcomb visited Seattle and signed on with Weyerhaeuser as a sawyer at its Snoqualmie Falls lumber mill. For more than forty years he worked in a variety of increasingly senior positions en route to Weyerhaeuser's top executive ranks.

From his early experiences, he developed an acute awareness of the outdoor world and the need for conserving it. These sentiments lasted a lifetime. He believed that nothing was to be hunted that wasn't intended to be used as food, nothing cut down or harvested that wouldn't be renewed. At Weyerhaeuser, he encouraged the company to promote itself as the "green" forest products company, emphasizing reforestation and the renewal of its natural resources.

In 1960, with retirement on the horizon and rheumatoid arthritis sapping his energy, he was planning the transition to a simpler life, one reminiscent of his rural boyhood years. Owning a ranch would be ideal. He hoped to use his business and outdoor skills in managing a self-sufficient agricultural operation.

Jon Titcomb's letter revealed another motivation for buying a ranch. He envisioned a sanctuary that would outlive him, a place that would delight and inspire his descendants long after he was gone. He understood that future generations would find it difficult to secure sizable tracts of undeveloped land, and he prized the privacy and peace of mind such acreage could provide. Ultimately, he wanted a ranch for the family, not only as an agricultural enterprise but also as a retreat — an island of tranquility in an increasingly supercharged world. Jon felt that Moccasin Lake Ranch and the Methow Valley was that special place.

Four generations of our family have now witnessed the evolution of the Methow Valley's culture from a rural, agricultural orientation to one catering primarily to tourism and recreation. Where cattle once grazed, houses now stand; where there once may have been a local creamery or butcher shop, there is now a supermarket or trinket shop. The area's largest employer is no longer an agricultural enterprise but a four-season resort. A generation ago, crews of twelve to fifteen men could be seen in our fields during haying season. Now, modern equipment has replaced all but one or two operators.

Much of this narrative may seem like ancient history to many of today's Methow Valley residents and visitors, who have never experienced cooking over a wood-burning stove or waiting for the telephone party line to become available. They have not seen the Methow Valley as it existed only fifty years ago in the middle of the twentieth century — a place where cattle, alfalfa, orchards, bands of sheep, and potato fields dominated the scene, and where ranch activities routinely included milking cows and butchering hogs or lambs. They will never experience the thrill of sledding down the snow-covered roads of Elbow Canyon in the dark of night, flashlights the only illumination except for the bonfire waiting at the bottom of the hill. Hopefully, this writing will bring to life a part of the Methow's past for them and help recent arrivals understand the valley's evolving culture.

The descendants of Jonathan and Gaye Titcomb — three children, six grandchildren, and sixteen great-grandchildren, as of 2004 — have taken a great interest in Moccasin Lake Ranch and the many activities available to them. Horseback riding, hiking, fishing, and camping at the lake in the summer and cross-country skiing, sledding, snowmobiling, and skating on the ponds in the winter all vie for their attention during the various seasons of the year. Originally, my purpose in writing this book was simply to set forth for them and their descendants a description of life at Moccasin Lake Ranch from its earliest days to the present.

I soon learned, however, that the evolution and development of the entire Methow Valley is closely aligned with that of our ranch. During my research, I often found it difficult to separate the individuals and events of Moccasin Lake Ranch from those of the rest of the Methow. As a result, I expanded the scope of this book to record the story of both Moccasin Lake Ranch and the beautiful valley where it lies.

Jon Titcomb's direct descendants (from left): Nim Titcomb, Gaye Pigott, and Greata Beatty. Moccasin Lake Ranch collection

Moccasin Lake Ranch from
Patterson Mountain.
Moccasin Lake Ranch collection

Chapter 1

Ancient Eden

As the earth's core cooled, tumultuous geologic activity shaped the Methow Valley and over millions of years gradually transformed it into a generous provider of food and shelter for its earliest inhabitants.

Under a hot summer sun, the hillsides of Moccasin Lake Ranch are a dry and unrelieved golden brown. Clumps of bunchgrass grow scattered among bitterbrush and sagebrush, but long gone are the wildflowers of spring. It seems incongruous that this parched land was once submerged under a great body of water, yet evidence of an ancient sea abounds. Marine fossils can still be found nearby at elevations almost fifteen hundred feet above the floor of the Methow Valley. Nearby, at the base of Patterson Mountain, a rock outcropping has been dated back some 150 million years, its sedimentary composition confirming our once maritime environment.

The valley's climate, topography, and combination of canyons, streams, terraced hillsides, and lakes give ample evidence of its active geologic history. A basic understanding of the processes that sculpted this richly endowed valley and influence its climate can add to an appreciation of its qualities. It also will help explain its attractiveness to Native Americans and to the white settlers who came later. In the following pages I have summarized some of the important geological developments that have molded the Methow Valley over the millennia.

The geologic activities that led to the formation of the North Cascade Mountain Range, bordering the Methow Valley on the west, can be traced back at least to the time of that Patterson Mountain outcrop. Two hundred million years ago, two great plates of the earth's crust, the Pacific plate on the west and the North American plate on the east, gradually edged toward one another. The two plates continually collided during a period of almost 100 million years at what was then the western edge of North America. The monumental collisions produced vast accumulations of mud and sand that built up the ocean's bottom. In the course of time, the denser Pacific plate slid under the North American plate, further raising the earth's crust.

About ninety-five million years ago the first vestiges of land appeared in the vicinity of today's Methow Valley. In the process, these geologic activities caused the shoreline of the Pacific Ocean to migrate westward almost two hundred miles. Extensive volcanic activity along the new margin of North America thrust great quantities of magma through the earth's crust and created the North Cascade

Mountain Range. During the years of tectonic collisions, the earth's crust hardened, thickened, and left the north-south fault lines just west of the Methow that trigger the earthquakes occasionally experienced in the Northwest today.

In addition to offering majestic splendor, the Cascade Mountains contribute significantly to the Methow's weather patterns. The range separates this inland valley from the moisture-laden air of the Pacific Ocean, causing the Methow to have drier air and greater temperature extremes than the coast. The result is that the Methow Valley has a semi-arid climate with short, hot summers, long, cold winters, and periodic droughts that last for several years. These weather patterns, as well as its topography, have long influenced human existence in the Methow Valley.

About one million years ago the Methow's landscape was dramatically reshaped by the continental glaciers that had formed in the north. These glaciers, which attained a thickness of over 4,500 feet, expanded southward from Canada, pushing enormous quantities of material before them. One legacy of this period is the Methow River's gravelly bottom, which makes it an inviting spawning site today for sea-run game fish such as salmon and steelhead.

More recently, starting about 110,000 years ago, the Cascade Mountains also spawned glaciers, known to geologists as alpine glaciers. The two glacier systems helped put the finishing touches on the Methow Valley's topography. Thousands of years after their formation, when the glaciers warmed and started to recede, they released torrents of water that carried sediment, boulders, and other matter along their course. These materials encircled blocks of ice in the path of the torrent until

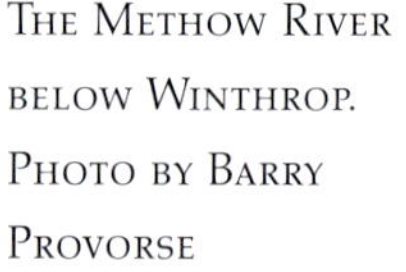

The Methow River below Winthrop. Photo by Barry Provorse

ten to twelve thousand years ago, when these ice monoliths gradually melted and became the lakes dotting the Methow today. Twin Lakes, at the foot of Patterson Mountain, are excellent examples of this process of lake formation.

Other waters from the melting glaciers relentlessly cut swaths through the valley's bottomlands. Alpine glaciers carved the ravines and canyons on either side of the valley, giving the Methow its distinctive features. The waters found their way to the Methow River and, eventually, to the mighty Columbia.

During the next several centuries, grasses, shrubs, and trees took hold in the soil deposited along the river and on the valley's stepped terraces. Vegetation provided cover and feed for a wide variety of animals. Bears, wolves, lynx, and cougars followed browsing deer and smaller mammals into the valley. Fish found their way up the Methow River to suitable spawning grounds during the spring and summer. Songbirds, hawks, and other birds also found the valley to their liking. Tall evergreens and thickets of chokeberry, serviceberry, and wild rose provided excellent nesting, while nuts and berries provided year-round feed. Migratory birds, including ducks and geese, became plentiful with the formation of lakes and ponds throughout the valley.

The first human migrants to North America are believed to have crossed some twenty-five thousand years ago over a land bridge that once connected Siberia and Alaska. The land bridge was created when arctic glaciers locked up so much of the ocean's water that its surface dropped hundreds of feet to expose the route. Prehistoric people gradually moved eastward, following the game animals that provided them food. As millennia passed, the hunters eventually migrated southward down the eastern slopes of the Rocky Mountains.

The first humans probably arrived in the Methow Valley about seven thousand years ago. Some anthropologists believe that humans may have come to the region even earlier—as early as eleven thousand years ago. They point to the discovery of spears and arrowheads in the East Wenatchee area dating to that time. These Clovis-era artifacts are the oldest evidence of human activity in the entire Columbia Plateau. However, as of the early twenty-first century, no evidence of that antiquity has been uncovered in the Methow Valley.

Experts believe that some forty-five archaeological sites can be found in the valley today. These sites attest to prehistoric human presence in the Methow Valley and to its desirability as a place to live in those times. Evidence of prehistoric man closest to Moccasin Lake Ranch includes camping sites alongside Twin Lakes, near Winthrop's Mack Lloyd Park, and at the mouth of Wolf Creek. These are places where the valley's earliest inhabitants lived over the winter months or camped during the gathering of roots and berries in the spring and summer. Not surprisingly, the sites are most commonly found near the mouths of creeks or alongside lakes where they afforded easy access to water.

At the Wolf Creek site, the outlines of four depressions are believed to be the remains of sweat lodges — precursors of the modern-day sauna. The sweat lodges consisted of pits dug into the ground and covered with roofs of tules woven together or skins supported by branches or logs. A nearby fire was used to heat rocks that early inhabitants sprinkled water on to create a steam bath for cleaning and purifying themselves. Much larger pits, also found in the Wolf Creek area, probably served as winter shelters. They were dug into the ground or banks, covered with skins, branches, and mud, and outfitted with a small opening in the roof to let smoke out. Near Wolf Creek, several burial sites of our valley's earliest inhabitants are still identifiable by their concentric stone rings on the ground.

The earliest inhabitants of the Methow Valley discovered that it was a generous provider of the essentials for human survival. They were readily able to find building materials, food, and pristine water. The geologic events that occurred long before their arrival had created a climate that was conducive to year-round survival while forming a physically imposing topography that limited access. This combination of circumstances perfectly suited the hunter-gatherer culture of early Methow inhabitants, permitting them to live unaffected and unmolested by the outside world. As a result, their lives changed little over thousands of years. Their customs and even their dialect were unique to the Methow.

Life for these people revolved around a seasonal cycle geared to gathering and preparing food. Each spring the women dug fresh supplies of roots: wild onions, Indian potatoes, bitterroot, desert parsley, and the most versatile staple of all, balsam roots. Summer was the time when men caught the fish — suckers, steelhead, and chinook and coho salmon — that may have constituted almost half of their annual diet. In September the people harvested wax currants, serviceberries, chokecherries, thimbleberries, and dogwood berries. Later autumn brought blueberries, elderberries, and pine nuts.

The most important time of year was the fall. Then the hunting season began in earnest. After the leaves fell from the trees, visibility improved in wooded areas, making it easier to find the bear and deer. Moreover, the animals were fat after a summer's feast of berries, grasses, or fish, and their hair had grown long in preparation for winter. Thus they provided not only more desirable meat and fat for cooking, but also better hides for shelter and warmth. After the men completed their hunts and the game was cleaned and stored, the Indians rested during the Methow winter, subsisting on the fruits of their labors and preparing for the next cycle of gathering and hunting.

Despite the Methow's isolation, tragedy struck one winter in the early 1780s. Smallpox killed almost two hundred members of the Methow's small native population. It's not known precisely how this European disease was introduced into the Methow, since it preceded the first white visitors by some thirty years. However, the disease was prevalent in a large portion of the U.S. West at that time. Perhaps some trapper or prospector visited the valley and unwittingly served as a carrier of the lethal germ. The epidemic's legacy was that many tribal members began leaving the Methow during winter for the more hospitable climate of the Columbia Basin.

The tribal members' winter migration to the Columbia Basin resulted in an important development that had long-term consequences for the Methow Indians. During the late 1700s they came into contact with horses for the first time. Using horses, tribal members found that they could greatly expand their hunting and trading territories, simplify their seasonal relocations, and enhance their social life. Riders competed in races, relays, and agility events as their horsemanship reached high standards. During ensuing winters, more Methow Indians began going south, where the climate was milder and feed for their horses was readily available.

PRE-STATEHOOD MAP OF WASHINGTON, 1879. UNIVERSITY OF WASHINGTON LIBRARIES, SPECIAL COLLECTIONS, UWM133

Sinkiuse-Columbia chief Moses poses for a portrait, Spokane Falls, Washington. Northwest Museum of Arts & Culture, Spokane, Washington

Chapter 2

When Worlds Collide

The importance of the Homestead Act of 1862 is hard to overstate. It is regarded as having done more to settle the West than any other single governmental action.

The Indians of the Methow likely saw their first white men in the valley during the summer of 1811. On the morning of July 6, villagers at the mouth of the Methow River were visited by David Thompson and his band of explorers. They were canoeing down the Columbia River in search of a waterborne trade route to the Pacific Ocean. Thompson reported that he exchanged gifts and smoked a ceremonial pipe with the "Smeethhowe" villagers. The Indians then provided his party with horses to portage around the Methow rapids (below the present-day site of the town of Pateros) before continuing down the Columbia.

Thompson was the first white man known to have seen the Methow River. Born in England in 1770 and orphaned at an early age, Thompson joined the Hudson's Bay Company when he was only fourteen and was sent to eastern Canada for training in the fur business. In 1797 he joined the North West Company, of Montreal, a competitor of his former company, and began exploring water routes to the Pacific. In 1810 the North West Company directed him to proceed down the Columbia, specifically to prevent the American-owned Pacific Fur Company from establishing a trading presence at the mouth of the river. After his brief stop at the Methow, Thompson continued down the Columbia and eventually reached the Pacific Ocean, only to find his American competition already ensconced on its shores at a settlement called Astoria.

In 1814, three years after David Thompson passed by the mouth of the Methow River, a hardy Scot named Alexander Ross trekked across the Methow Valley in search of an overland route to the Pacific coast. Ross, age thirty, was the second white man known to have visited the Methow, and the first to enter its upper valley. He had started his career as a schoolteacher before emigrating to eastern Canada. Ross went to work for the North West Company, which directed him to attempt a crossing of the Cascade Range. The goal was to determine whether an overland path could provide a shorter, faster fur-trading route to the Pacific than the trip by boat down the Columbia to Astoria.

Ross, who was apprehensive by nature, became suspicious of the local Indians soon after his arrival in the Northwest. He thought them to be lazy. Nonetheless, on July 25, 1814, Ross and three Indian guides — members of the Okanogan tribe — forded the Methow River near its confluence with the Columbia, climbed past Alta Lake, and began what turned into a tortuous journey over the mountains. The specifics of their route are not clear, but it appears that they traveled far up the Methow Valley to the Twisp River, then turned west toward the North Cascades' tangle of glaciated peaks.

From the Twisp River, the party most likely proceeded either over War Creek Pass to Stehekin and then over Cascade Pass to the Skagit River or over Twisp Pass to Granite Creek and on to the Skagit. One of his Native guides fell ill during the crossing, and Ross assigned a second guide to remain behind with the man. The third and last guide turned back in the face of snowstorms and high winds, leaving Ross to proceed alone through dense forest and steep canyons. After several more days he arrived at the upper reaches of the Skagit River.

At the end of his physical limits, Ross turned back — not realizing he was then within a few days of relatively easy travel to the ocean. He returned to his company's trading post on the Okanogan River after having been gone for less than one month. Ross wrote that this journey was the most arduous he had ever undertaken. His notes carried strong recommendations against any further exploration of an overland route through the Methow, commenting that the mountains were treacherous. To reinforce his conclusions, Ross also reported that there were not enough Indians in the territory to make trading viable.

Thompson and Ross were among the first explorers to visit the Pacific Northwest during the early years of the nineteenth century. Trading companies coveted this area because it was a rich source of sea otter and beaver pelts. But by the mid-1830s, beavers were being trapped at an unsustainable level, and along the Pacific coast, sea otters had virtually disappeared. By mid-century, after establishment of the Canadian – U.S. boundary at the 49th parallel, the British had pulled out of American territory. The era of corporate trapping had come to an end, leaving this occupation to the rugged individualists who lived by their wits and instincts in the wilds of the mountainous West.

In the decades following the visits by Thompson and Ross, only a few white trappers and prospectors made forays into the Methow Valley. However, important developments were taking place in the nation's capital that would ultimately impact the Methow's future. By the middle of the nineteenth century, the U.S. Congress was encouraging expansion into the nation's western territories. Settlement of this

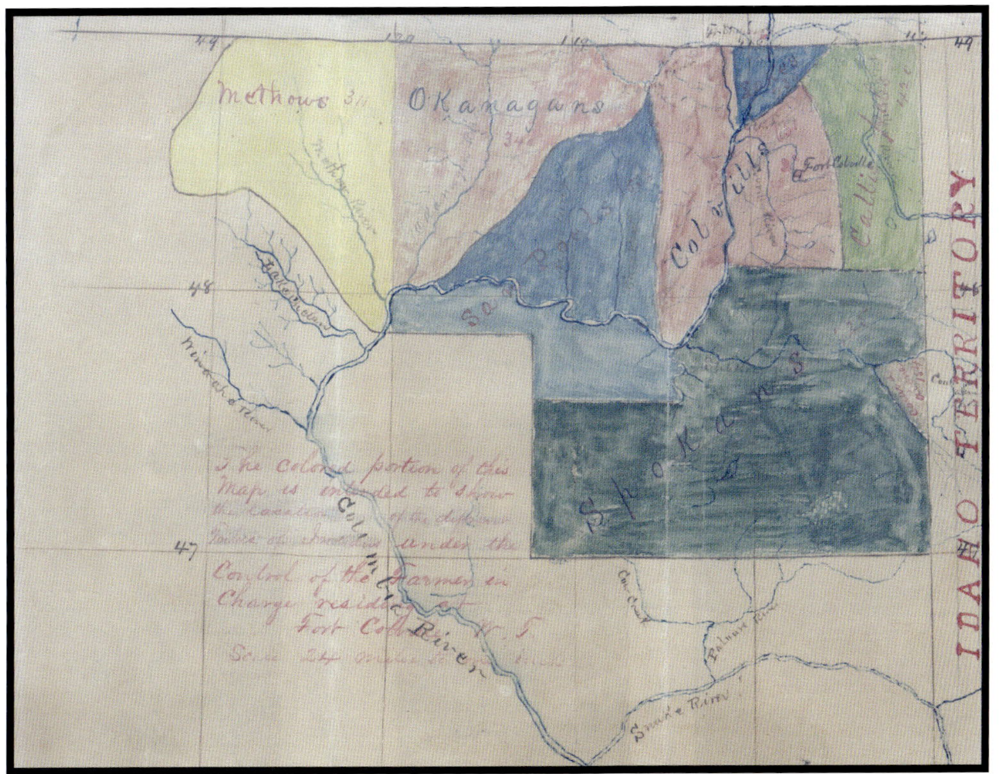

Sub-Indian Agent William Parkhurst Winans's hand-drawn map delineated the area and number of tribes he served between 1870 and 1872. Manuscripts, Archives and Special Collections, Washington State University Libraries, WSU 133

largely unknown and unmapped land would serve the national interest. Settlement would discourage incursions by foreign governments and help to protect settlers from sometimes hostile natives. Although it was preoccupied with the Civil War, Congress passed the Homestead Act, and it was signed into law by President Abraham Lincoln in 1862. This act is often credited with doing more to settle the western United States than any other piece of legislation.

The Homestead Act permitted anyone to obtain ownership of 160 acres of land in the public domain as long as the person agreed to live on the homestead for at least five years and make certain minimal improvements. Additional governmental inducements were made to railroad companies in the form of land grants. Congress wanted transcontinental railways built to improve the reliability, speed, and safety of cross-country travel. These inducements were obviously successful in stimulating large-scale migrations across the country. In the 1870s, some 250,000 homestead claims were filed. In the 1880s, that number almost doubled. During the first four decades of homesteading, almost 1.5 million pioneers made claims to 270 million acres of U.S. territory.

As early as the 1850s, the U.S. government, recognizing the potential for serious conflict between natives and whites, began studying various solutions. The farming lifestyle and territorial possessiveness of the settlers were in direct contrast to the nomadic ways of the natives. The Columbia Basin, which encompasses the Methow Valley and much of what is now Eastern Washington, was sparsely inhabited in the 1870s. Only a dozen Indian tribes, comprising fewer than five thousand Indians, hunted in the basin, while only a few hundred white families were randomly located throughout the area. Nonetheless, as the Columbia Basin became more populated, reports of increasing friction between white settlers and Indians filtered back to the nation's capital.

To address this problem, the government decided to set aside portions of the public domain exclusively for Indian use. Originally, treaties were negotiated with several Indian nations. However, in 1871, Congress unilaterally determined that it would no longer regard native tribes as separate nations. From that time on, there was no need for congressional negotiations with a multitude of tribes. A U.S. Bureau of Indian Affairs was established and decisions on land use were henceforth determined by presidential order. This development gave U.S. presidents great latitude in specifying the size and location of lands to be set aside for tribal use. Perhaps more important, the boundaries of the Indian reservations could be aligned or moved to serve the interests of newly arrived white settlers.

Not surprisingly, as the number of white settlers and fortune-seekers increased in the West, and as its great natural resources were discovered, local politicians and vociferous miners encouraged the president to shrink the size of the reservations and to open ever more land to the settlers. In April 1872, President Ulysses S. Grant signed an executive order creating the original Colville Indian Reservation within the area then known as Washington Territory. This order set aside for Indian use thousands of acres of what is now northeastern Washington. It included all lands east of the Columbia River to the Pend Oreille River and from the 49th parallel south to the Spokane River.

Three months later, after receiving an urgent petition signed by the government's Indian agent and by sixty settlers (half with Indian wives) whose land would have been taken from them, Grant returned all of this original real estate to the public domain. The president then replaced it with a second reserve — also known as the Colville Reservation — that was located just to the west of the original. This land had the same northern and southern boundaries but lay between the Columbia River on the east and the Okanogan River to the west. It did not include the Methow Valley, which remained in the public domain but was unsettled because of its remoteness.

In 1879, President Rutherford B. Hayes signed an order creating another reserve adjacent to the relocated Colville Reservation. This was formally known as the Columbia Reservation but was often referred to as the Moses Reservation, in recognition of Chief Moses. This six-foot-tall, two-hundred-pound warrior was generally recognized by several local tribes as their chief. The original Moses Reservation included all lands between the Okanogan River on the east and the crest of the Cascade Mountains to the west, from the 49th parallel on the north to Monse, located at the confluence of the Columbia and the Okanogan Rivers. From this point, at the southeast corner of the reserve, the boundary line turned west through Malott and back to the Cascade Mountains. Only a portion of the Methow Valley was included in this new reserve.

The boundaries of the Moses Reservation also were soon modified, this time in response to the demands of Chief Moses, who wanted more territory included. In 1880, the entire Methow Valley became part of the reservation. However, the boundaries continued to change. In 1883, a fifteen-mile strip of land lying south of the 49th parallel was removed from the reservation at the strong request of mining interests. In 1884, President Chester Arthur stepped in with an order that withdrew all lands west of the Okanogan River from the reservation and returned them to the public domain — a decree that in effect eliminated the Moses Reservation altogether. President Arthur's order made all of the Methow Valley available to white settlement as of May 1, 1886.

The year 1886 can be regarded as the beginning of the modern era in the Methow Valley. Until that time the valley's only inhabitants had been indigenous bands of Indians and occasional trappers and miners. The trappers and prospectors spread the word that the Methow possessed ideal conditions for settlement: vast stands of bunchgrass for livestock, timber for building shelters, and an abundance of berries, fish, and game for food. Water from the Methow River and its tributaries was pristine, and the climate was agreeable. Because the land had been Indian territory, the soil had not been plowed or disturbed. All of these virtues would soon catch the attention of enterprising families looking for a better life. The situation was changing fast as government policy, aided by the newly built railroads, brought relentless settlement of the western states.

University of Washington mining students at Red Shirt Mine, 1905. University of Washington Libraries, Special Collections, UW 6812

CHAPTER 3

The Lure of Gold

Hopeful prospectors trekked into the Methow after hearing reports of gold discoveries from Squaw Creek in the south to the Slate Creek district in the north.

The search for gold brought the earliest settlers to the Methow Valley. As fur trapping declined in the first half of the nineteenth century, prospecting for precious minerals picked up. By the mid-1800s, copper, silver, zinc, and especially gold had been discovered in many of the western mountain ranges. Gold strikes were powerful inducements for young men to venture west.

The earliest sustained prospecting in the Methow came in the late 1850s or 1860s, mainly by Chinese immigrants brought to the United States to help build the transcontinental railroads. Once that work was largely completed, these industrious workers began panning and sluicing for gold along the Columbia River and its tributaries. These techniques — known as placer, or surface, mining — had the distinct advantage of requiring little capital investment. Miners who panned for gold could literally carry all the equipment they needed on their backs. Once they found water and gold-bearing sand or gravel, they were ready to go to work. A little agitation of the pan as it held sand and water was enough to separate the heavier gold particles from lighter materials.

Although panning was the simplest and most widespread placer-mining method, it was slow and limiting. A more productive method was to run water through a series of sluice boxes. These boxes were often placed end-to-end and could extend to a length of 150 feet or more. The flowing water propelled ore-laden sand and gravel along the troughs. The heavier, gold-bearing material was trapped against riffles, or crossbars, fixed to the bottom of the troughs. This material was then collected, and panned if necessary, for a final separation of the gold.

In the 1860s and 1870s, Chinese placer miners constructed a five-mile ditch along the lower Methow River to enhance their efforts to find the elusive gold nuggets. This so-called China Ditch intercepted water at its headgate about three miles up the Methow River and carried it to the north bank of the Columbia River and then almost a mile downstream from today's town of Pateros. At various points along the ditch, the miners built sluice boxes and flumes using logs they found along the river. They diverted water from the ditch into these sluice boxes to begin the separation process. Leftover material that had been sluiced along the ditch formed great gravel bars that could be seen on the Columbia's banks until they were flooded almost one hundred years later by backwaters that were created by the river's hydroelectric dams.

The China Ditch was capable of carrying huge quantities of water. After it had no further use in mining, it was used to carry water to orchards along the Columbia nearly to the Chelan County border. The orchardists in the area used its water for irrigation until the great flood of 1948 destroyed the headgate and sections of flume.

In time it became apparent that the ore at the mouth of the Methow River was of low grade and limited in quantity. After hearing reports of gold strikes in the Fraser Valley of British Columbia, many fortune hunters left to seek riches there. The diligent Chinese stayed at the mouth of the Methow into the 1870s, finding sufficient gold not only to sustain themselves but also to send money to family members in China. Nonetheless, the exodus to the north did turn out to be a harbinger of future mining disappointments in the Methow—ventures that started with promise but ended in disillusion.

Hard-rock, or lode, mining in the Methow dates to the mid-1880s. Unlike placer mining, hard-rock mining often requires a substantial investment in equipment and mine development. It also is more dangerous, frequently requiring the use of explosives to blast shafts and tunnels.

Hard-rock miners in the Methow searched for promising outcroppings or veins containing ore. There they would burrow underground to look for seams, or lodes, holding precious minerals. Material that looked promising was transported by wagon, and later by aerial tram, to concentrating mills, where jaw crushers and mechanical sledgehammers (known as stamps) or other pulverizing equipment broke the large rocks into smaller pieces. Sometimes the material was broken into still smaller pieces by grinding machinery (such as arrastres) to further the extraction process. Many concentrating mills incorporated a flotation process using chemicals such as cyanide or mercury to remove impurities. After the concentrating process, the ore was shipped to smelters for refining into marketable bars. These smelters were located outside the Methow, as the local mines never produced enough concentrate to justify building a smelter in the valley. Most concentrate from the Methow was freighted to Tacoma or Everett in Washington or to Nelson or Trail in British Columbia.

One of the first known attempts at lode mining in the Methow was carried out by a young man named Dick Miller. Miller started out as a teenager from Oregon Territory in 1880 and headed for the Methow's Slate Creek district, at the far northwest end of the valley. It is said that he was lured to the valley by rumors of an earlier mineral discovery. As the story goes, Miller did discover a gold-bearing vein — but after he left the site in search of food, he couldn't find it again. After a summer of prospecting without further success, Miller returned empty-handed to Oregon. He reappeared in the Methow in the early 1890s with his new wife and moved onto a farm near Twin Lakes, at the foot of the property that later became Moccasin Lake Ranch. But Miller's original instincts for gold were right on target: by the time he and his wife had returned to the Methow, the famous Eureka lode in the Slate Creek district was offering up its riches.

John "Chickamun" Stone, another early Methow prospector, began searching around Beaver Creek in 1885, southeast of the eventual site of Twisp. Local Indians gave Stone his colorful nickname, which means "money" in their tongue, apparently believing he had a special affinity for financial success. However, subsequent events may have proved them unduly generous in their assessment.

Stone is generally credited with making the Methow's first gold strike that resulted in commercial development. This occurred in 1887, in Findley Canyon near Polepick Mountain. This strike led to creation of the Red Shirt Mine, which produced both gold and silver until near the end of the century. By 1897 the mine employed a crew of fifty, making it one of the valley's major employers. Unfortunately for Stone, by the time the mine was in peak production, he had long since sold it to James M. Byrnes. The price paid by Byrnes was a red shirt and a bottle of whiskey. Byrnes went on to operate the Red Shirt Mine and to run both the local trading post and the post office at Silver. Inclusion of whiskey as part of the purchase price may have been prescient because the generally accepted view today is that the mine was later used by Buck and Boots Lewis to house the still that they allegedly used to make moonshine whiskey during Prohibition.

While prospecting at the Red Shirt was getting under way, mining activity in the lower valley was also generating excitement. On Squaw Creek, several mines bearing such names as Ocean Wave, Highland Light, Philadelphia, Paymaster, Hidden Treasure, and Excelsior were producing gold as well as impressive quantities of promotional materials. The latter proclaimed unlimited prospects for anyone willing to join in. The advertisements must have been convincing, as they generated a mining boomlet in the mid-1890s that included the employment of about three hundred miners. However, production began to drop off around 1897. Shortly thereafter, word began to spread about a fabulous gold discovery in the Yukon. It wasn't long before the little community, originally known as Squaw Creek but later renamed Methow, was abandoned. This first town in the valley called Methow turned into a ghost town before the turn of the century.

A man named Gilbert was prospecting in the early 1890s up the Twisp River. He filed that district's first mineral claim after finding gold there while working alone.

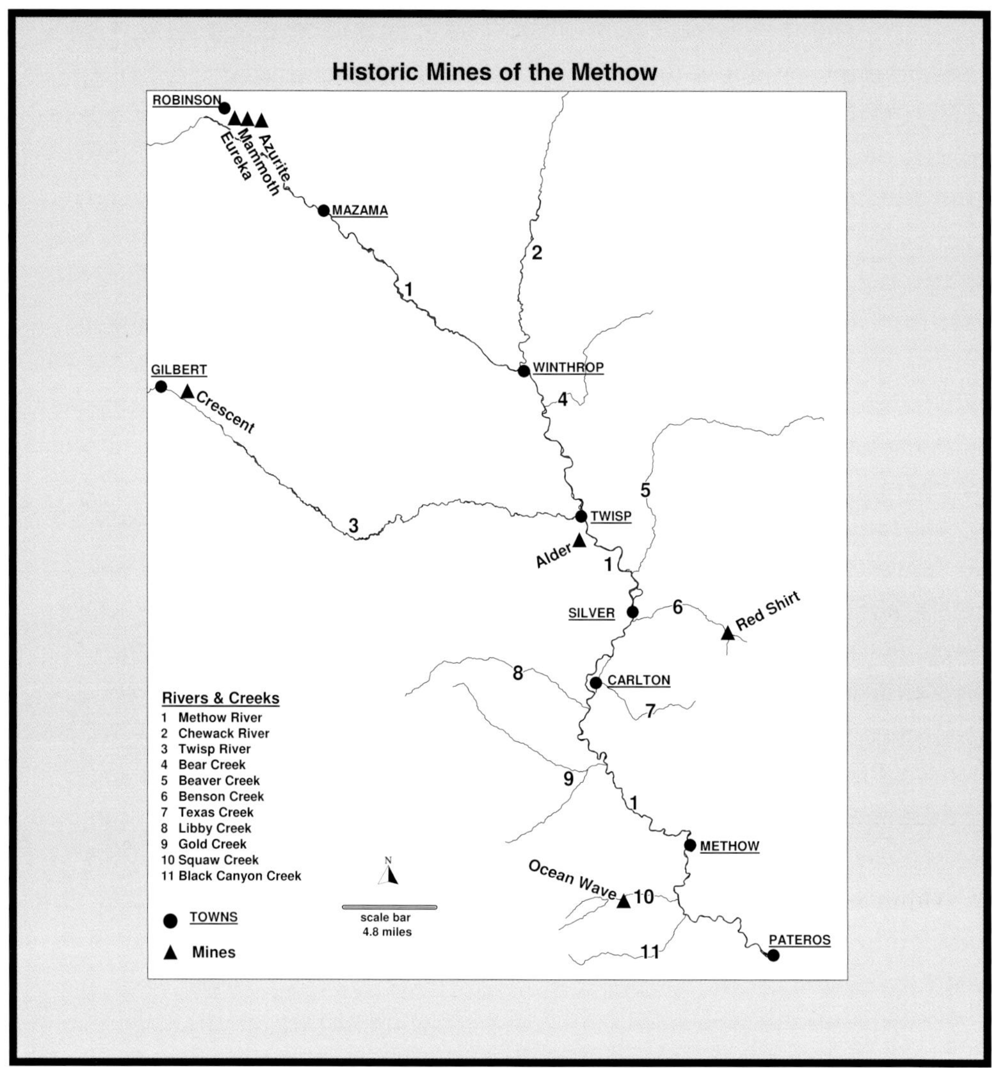

Gilbert discovered enough valuable ore to personally finance a wagon road from his claim down to the Methow River at the future site of Twisp. The road and Gilbert's strike helped start a small mining boom that, by the mid-1890s, grew to more than two hundred claims scattered along the tributaries of the Twisp River. Gilbert was later injured in a mining explosion and never returned to the valley. The untimely accident cost Gilbert his hand, but his name remains on the tiny settlement and towering mountain near the end of the Twisp River Road.

Another pre-1900 discovery of gold in the Methow occurred in 1896. A few miles southwest of Twisp, a site that is now known as the Alder Mine was discovered by T. H. Culbertson on McClure Mountain. Because its remote location in the Cascade foothills, at an elevation of 3,600 feet, made access to the site difficult, the property was only sketchily explored for several years. Almost three decades of limited development followed the initial discovery of gold traces. Then in 1939, the Alder Group Mining Company, through its lessee, the Methow Gold Company, produced enough concentrate to ship two railcar loads of material each day to smelters on the west side of the Cascade Mountains. During the 1940s, the Alder Mine continued to produce at full capacity. The mine eventually became the state's all-time leading producer of precious minerals, generating over one million dollars' worth of gold, zinc, silver, and copper. But after World War II the price of zinc and silver fell to a level that made the mine's low-grade ore uncompetitive, and it was closed in 1953. In 2003 the mill's timbers and planking were finally being removed, reducing the mine to only a memory.

Charles H. Ballard, M.E., president, Azurite Gold Company, Twisp, Washington. Shafer Museum Collection

In March of 1895, Colonel W. Thomas Hart, a flamboyant and well-financed agent for Montana mining money, arrived in the valley. Hart proclaimed loudly and widely that he had taken options on a mine in the Slate Creek district, the region originally visited by Dick Miller. More recently the area had been made famous by Alex Barron's discovery of the Glory Hole at the Eureka Mine in 1893 and by further discoveries a year later. Hart planned to build a wagon road up the north side of the Methow River. It was to go past Lost River and on across the Cascade summit. The summit would be crossed at Slate Pass (later called Harts Pass), which at an elevation of 6,197 feet would be the state's highest.

Within a month Hart had sixty-five men working on the project, saying he was prepared to spend $100,000. The undertaking was a godsend for the people of the Methow Valley. It provided severely needed employment and infused a great deal of new money into the local economy.

To accomplish this formidable project, Hart hired Charles Ballard, a widely respected civil engineer and mineral surveyor from Oregon. Ballard's long career eventually involved every aspect of mining. He had even laid out the mining towns of Ruby and Conconully in the Okanogan region during their mining heyday a decade before. A major Cascades mountain above the Azurite Mine in the Slate Creek district was later named for Charles Ballard.

Hart made Ballard responsible for turning the existing packhorse trail into a usable wagon road. But just two months after starting the road-building, Hart abruptly paid off his men and stopped all further work. The speculation was that he had used up his money. Hart departed in disgrace, reportedly for Arizona, and never returned to the Methow. Scarcely six years later, in 1901, he boarded a freight train in Idaho and, as it passed over a high mountain bridge, jumped to his death in the surging waters of the Snake River.

The wagon road was completed by Charles Ballard and his brother, Hazard. After crossing over the Cascade summit, it proceeded about three miles down the west side of the divide to the mining camps. During the district's peak years of 1897 and 1898, Alex Barron's Eureka Mine, the Ballard brothers' Azurite Mine, and Charles Ballard's Mammoth Mine all produced gold, silver, and some copper.

The combination of wagon-road building, mine development, freighting of supplies and machinery, outfitting of prospectors, and building of hostelries along the route brought the early Methow settlers to a state of great anticipation of a prosperous future. The way stations of Ventura, Robinson, and Chancellor on the Methow (east) side of the Cascade crest as well as the town of Barron on the west side were created to support the hordes of men needed to work and service the mines. At Barron the men could find saloons, dancing girls, a butcher shop, and a trading post, plus a post office where mail was received by ski, dogsled, or horseback, depending on the season.

At its peak this territory employed as many as twenty-five hundred workers — many of them rowdy, rawboned, and mustached characters known for their drinking and swearing. It is said that one such character, Arthur Egbert, had a standing bet of one hundred dollars that he could outwalk a horse to Bellingham or to Ives Landing (Pateros). Apparently he was never challenged. Food at the camps was hearty and plentiful. Cattle were sometimes driven from the Methow as far as Barron, where they were butchered for local consumption.

Those heady Slate Creek district mining days were brief. The mineral content of the readily obtainable ore began to diminish, while operating costs remained sky-high. Miners began to look elsewhere. The call of the Yukon and its Klondike gold discovery was being heard loud and clear by miners in the Methow's northern district just as it was heard by those at Squaw Creek in the south. These men yearned for the excitement of staking their own claims. Soon after the turn of the century, many of them abandoned the Methow and headed to the far north. Production at the Methow's mines was cut back or discontinued altogether.

Through the early twentieth century, mining activities were limited primarily to sporadic efforts by individual prospectors. A few corporate mining ventures arose during the 1930s in response to the nation's increasing mineral needs and improvements in local transportation. The Alder Mine near Twisp was by far the most productive. To the north, the American Smelting and Refining Company (ASARCO) leased a group of claims and five mills in the Slate Creek district from Charles Ballard in 1934. These never reached the potential that the company had envisioned, however, and the lease was allowed to lapse in 1940. Most of the equipment was removed, with some of it becoming part of the displays at the Shafer Historical Museum in Winthrop. Abandonment of these mines represented the end of mining in the Methow. A period of almost one hundred years of hope and toil in the valley forever closed. Investments of outside money in the Methow's mines were made at only sporadic intervals, but they provided a crucial economic stimulus to the valley at times when it was most needed. During pioneer days and again in the mid-twentieth century, the mines provided jobs at good wages. The services and construction projects required to keep the mines running contributed importantly to the local economy as the miners' paychecks were cashed and spent locally. But for investors in the mining properties, many dreams were shattered by glittering prospects that never materialized.

Barron, Washington, 1900. University of Washington Libraries, Special Collections, UW 2318

A. C. Libby family homestead on Libby Creek. Dick Webb collection

Chapter 4

The First Settlers

Pioneer families came from around the world to homestead in the Methow, a valley so remote that government surveyors had yet to complete their work.

There was no immediate stampede of pioneers into the Methow after the presidential order of 1886 opened the land to white settlement. The forerunners of settlers in the valley were the Robinson brothers, Jim and Tom. The Robinsons, like explorer David Thompson before them, grew up as orphans before migrating west. They had run a sawmill in Michigan that was twice burned to the ground by vandals. Left penniless, the brothers then trapped muskrats in the local swamps for eight cents apiece to earn their rail fare west. They traveled to Ellensburg, Washington, where they met their cousin Dick Jones and a friend of his, Shorty Brant. The young brothers — Jim was twenty-one, and Tom twenty-six — heard stimulating tales about the abundance of beavers in the Methow Valley and decided to try their hand at trapping these valuable animals.

The brothers had a total of eighty-five cents between them when they arrived at Ellensburg. They traded their travel trunk to a photographer for two old guns and a few dollars. Wearing the rubber boots they had used in the Michigan swamps and accompanied by Dick and Shorty, they began a trek of some 150 miles to the Methow.

The Robinsons had been told that local Indians would help ferry them and their gear across the Columbia River at Chelan. Arriving at the little settlement but seeing no activity, the group whooped and hollered to attract attention. Soon they saw about thirty-five naked warriors, who let out several war whoops of their own while staring menacingly at them. Knowing they could not safely retreat, the travelers mounted their packhorses and proceeded to ford the river directly toward the natives. After what seemed like a long time during which no words or signs came from the natives, their leader, Chief Long Jim, spoke to his warriors and broke the tension. Long Jim then pointed the Robinson group in the direction of their next destination. As Jim Robinson wrote, "if it hadn't been for Long Jim we would never have gotten across."

After days of rugged hiking, the Robinson brothers' rubber boots wore out. Jim and Tom wrapped their bruised and bleeding feet in gunnysacks and pushed on along the north bank of the Columbia River. At the mouth of the Methow, the group hired an Indian guide, Narcisse, to lead them into the valley. Narcisse loaded provisions onto his pack pony and supplied saddle horses for the others to ride over the Okanogan highlands into the Methow.

In late October 1886, the men descended along Benson Creek to its confluence with the Methow. They camped briefly in the nearby Beaver Creek area with prospectors Chickamun Stone and Jack Forrester. These men told the Robinson group that the beavers had mostly been trapped out in the northern part of the valley. Jim and Tom were suspicious that Stone and Forrester were misleading them in order to protect the northern Methow for their exclusive use. They quickly resumed their trip, settling for the winter near present-day Winthrop. There they built a 14-by-14-foot cabin out of cottonwood poles, which they covered with bulrushes or cattails. A four-foot-square opening in the roof provided smoke ventilation.

During that winter of 1886 – 87, there were only nine white men in the entire Methow Valley: the Robinson party of four; prospectors Stone and Forrester; trapper Jim Glover, who later homesteaded at the site of Twisp; mountain man and prospector Al Smith, who joined the Robinson group for the winter; and a fellow named Charley Mansur, who was ailing and soon died. Being new to the wilds, the Robinsons, not surprisingly, ran out of ammunition and most of their supplies before the winter had ended. They existed on a diet of beaver meat and salt for six weeks. In the spring, the brothers and Shorty Brant trekked back to Ellensburg. The Robinsons wanted to earn money to buy "a winter's grubstake, a new rifle, a six-shooter, and plenty of ammunition" for their stay in the Methow the following winter. But Shorty was finished. He never returned to the valley.

Returning to the Methow in the fall of 1887, the brothers discovered that their Winthrop camp had been located precisely where hundreds of Methow Indians assembled each spring to gather balsam roots. This prompted the brothers, along with three other men, to move considerably farther up the valley to the stream that eventually became known as Robinson Creek. The men quickly made the gloomy discovery, however, that most of the Methow's beavers had already been trapped. The humorous nicknames for their two camps — Camp Troublesome and Camp Hasty — suggest they realized that earlier reports of an abundance of beavers were grossly out-of-date. Stone and Forrester had been right. The Robinson group reported that their rations for the winter consisted of what they could shoot — deer and goats, sometimes ducks — plus bacon, potatoes, and onions. They occasionally trapped a wolverine to add variety to their diet. They also fried apple pie and doughnuts in a skillet, using bear oil or goat tallow.

By the end of their second winter in the Methow, the brothers and their friends, tired of daily treks to empty traps, had decided to try something else. Dick Jones had contracted tuberculosis and left the valley, and Al Smith had moved to the Okanogan Valley to take up storekeeping. Tom Robinson turned his hand to prospecting for minerals in the northern Methow Valley but enjoyed only modest success. He stayed in the valley until his death in 1926.

Jim Robinson established an apple orchard across the river from the present town of Methow and became the first Methow settler to experiment with irrigation. As the years passed, Jim helped develop an irrigation cooperative, ran sawmills, and served as Pateros postmaster. Robinson and his wife eventually moved to Oroville, in the upper Okanogan Valley, where they remained until Jim's death in 1939.

The Robinsons were prototypes of the pioneers whose versatility, energy, and vision settled the wilds of north-central Washington. In 1887, the year after the brothers' first trek into the Methow, a trickle of home-seekers ventured from Ellensburg to investigate the valley's potential. These men were not prospectors or trappers. Their lives were focused on farming and raising cattle. They were drawn to the Methow Valley by the reports of abundant bunchgrass, clean water, and rich, undisturbed soil.

In the vanguard were a group of friends and transplants from Wise County, Texas. They had worked neighboring farms and shared a mutual interest in leaving the South. The Civil War had been settled two decades earlier, but harsh memories persisted. Confederate money had become worthless; family savings accumulated before the war had been converted to Confederate coin and had no value. To make matters worse, farmers were confronted with a prolonged drought in the Texas Panhandle. When these young families heard about the Homestead Act and learned that, in the vast regions of the West, they could own land with agricultural potential they were eager to relocate and try their luck. A number of them crossed the Great Plains by wagon train in the mid-1880s and took up farming in Kittitas County near Ellensburg.

One morning in late April of 1887, four of these Texas transplants left their Ellensburg-area farms to inspect the Methow Valley. Mason Thurlow, Harvey Nickell, Napoleon Stone, and Robert Prewitt had found that the Kittitas Valley did not meet their farming needs. They thrilled, however, to stories about the Methow Valley's bounty. Because they were traveling by wagon, the men took a gentler and hence more suitable route into the Methow than that taken by the Robinsons, who had traveled on foot and horseback. The Texas group proceeded to the Big Bend of the Columbia River, where they crossed on a ferry that was little more than a collection of logs strapped together. Their horses swam across the river tethered to the raft.

Once on the north shore, the men followed the Columbia to the Okanogan River, which they forded at Monse. For the difficult trip over the Okanogan highlands, they abandoned their wagons and used packhorses to carry their supplies. At the site of today's Malott, the men stayed overnight at the L. C. Malott cabin. Their hostess, Mary Malott, is said to have told them that she had not seen a white woman for six months. She was therefore eager to encourage the travelers to settle with their families in the Okanogan. The weather was better than in the Methow Valley, she exclaimed. But the men resumed their journey, motivated by their curiosity about the Methow. Once at the Chiliwist summit, they picked up an old Indian trail and followed it to the headwaters of Benson Creek and on to its confluence with the Methow River.

As they descended to the valley floor, the travelers were awed by the spectacle before them. The clear waters of the rushing river, the tall bunchgrass, and the abundant game were all there, just as they had been led to believe. The men found serviceberries, chokecherries, and Oregon grape as well as herbs, roots, and wild parsley. They caught fish in the river. The bright blue skies seemed to confirm their most fervent hopes about the Methow's suitability for homesteading.

The Texans explored the upper portion of the valley for almost two weeks, going north as far as Boulder Creek, then crossing the Chewack River and traveling west to the Methow River. Their survey then took them south through the area of present-day Winthrop. They were favorably impressed by the abundance of bunchgrass that grew virtually everywhere. In addition, the men sought property that contained groves of trees for building materials and shade, open land for gardens and crops, and convenient access to flowing water. Each man soon found what he was looking for, and the group returned to Ellensburg to happily plan the move to a new home the following spring.

The summer of 1887 saw a few other adventurous pioneers enter the Methow Valley. Blizzards, drought, and other trials elsewhere in the country, plus a common lust for adventure, brought these men, some with families, to the valley. Included in this first wave of immigrants to the Methow were George Thompson, Jimmy Sullivan, John Hartle, Ben Pearrygin, Joe Frazer, Cap Wright, and Dan Bamber. Many Methow prominences have been named for them, as outlined in the brief commentaries that follow.

George Thompson, a bachelor born in eastern Canada, traveled the United States as a hardware salesman before becoming the first white settler on what is now Moccasin Lake Ranch. He settled along a small stream that bisects the ranch's bottomland and runs into the wetlands near the present-day rodeo grounds. This stream was named Thompson Creek in his honor. A prominent ridge that overlooks the valley from its western flank was also named for him. Thompson became

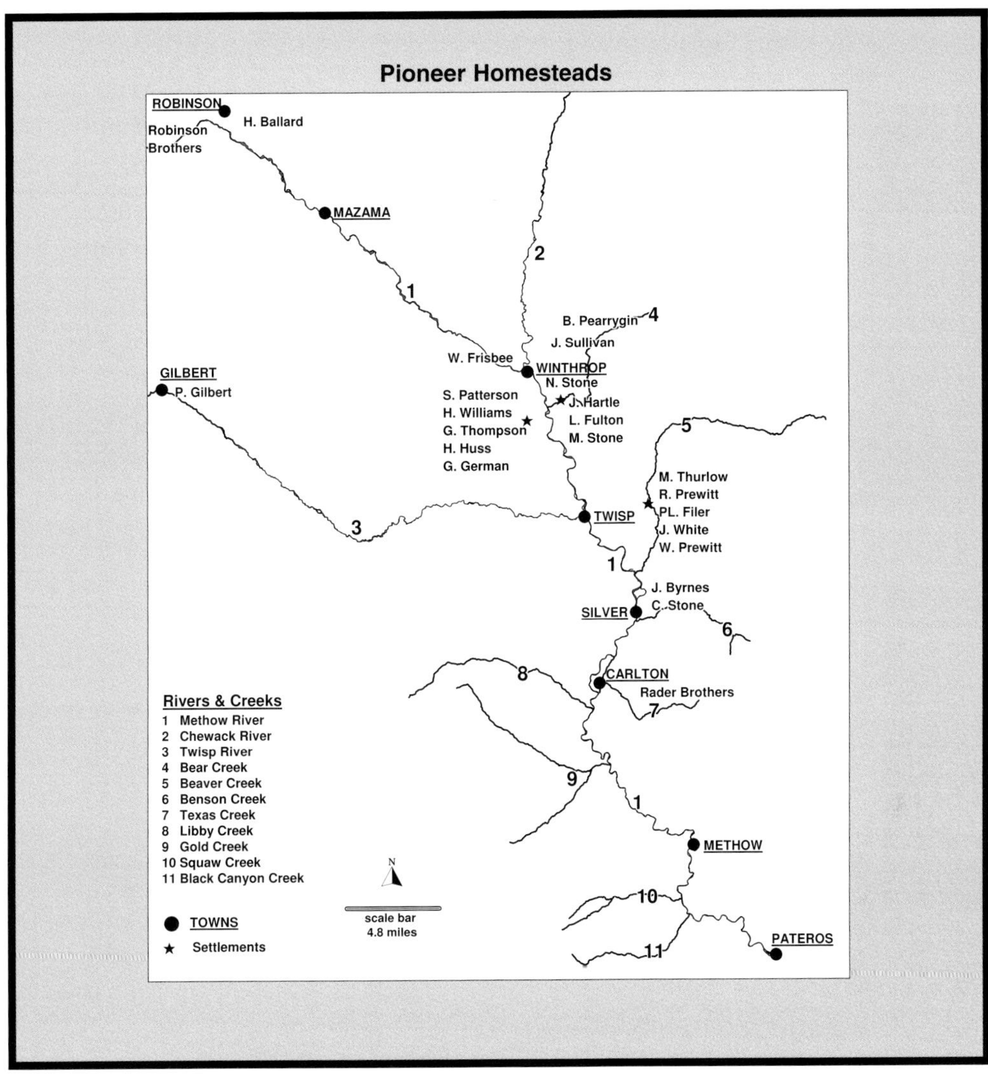

known in the valley for his oratory, his generous participation in community affairs, and his farming prowess.

New Yorker Jimmy Sullivan, a twice-wounded veteran of the Civil War, settled on the east side of the Methow River in midvalley. This colorful Irishman married divorcée Louisia Heckendorn and they later opened the first hotel in Winthrop, a simple log structure by a freshwater spring. Sullivan donated part of his homestead for use as the local cemetery. It bears his name today. Not far away is secluded Sullivan Pond, also named after him. Weakened by his war wounds, Sullivan died in 1900.

John Hartle, a widower with five children, was part of the Wise County, Texas, group that traveled to the Northwest by wagon train. Also a Civil War veteran, Hartle farmed near Ellensburg before venturing to the Methow. He built the first cabin near the mouth of Bear Creek and lived there with a friend, Tony Cogswell, during the mild winter of 1887–88. The following summer, Hartle returned to Ellensburg to harvest his crops. Returning to the Methow in the fall, Hartle built another cabin, below today's Bear Creek Golf Course. He invited his nephew, Lee Fulton, and Fulton's family to live in the original dwelling. A few years later Hartle sold his claim and cabin to Fulton for two milk cows and moved to the west side of the river, not far from today's Moccasin Lake Ranch. There he would be closer to his oldest daughter, Elsie, and her husband, Sam Patterson.

Ben Pearrygin lived on the shores of what became known as Pearrygin Lake. His name was also given to a state park, a creek, and a mountain in the Okanogan highlands. Joe Frazer staked his claim just upstream from the confluence of Beaver Creek and the creek that now bears his name. Dan Bamber settled that same year on what is now called Bamber Flats. Cap Wright Hill, near Balky Hill, commemorates Cap Wright's arrival in the valley in 1887.

Immigration into the Methow Valley continued to pick up the following year. The spring and summer of 1888 brought about twenty-five new families to the valley. Some of the transplants from Wise County, Texas, returned that spring from Ellensburg, this time bringing their possessions and their families. A four-wagon caravan left Ellensburg in early March of 1888, which gave the settlers enough time to complete the arduous journey, build shelters, and plant gardens before winter set in. Mason Thurlow and his son Will traveled with one wagon; another was used by Robert Prewitt and his brother William. The third wagon was taken by Peter Filer, Manford Stone, and Stone's daughter, Bertie; the fourth was that of Jewett Davis and Wally Maddox. Stone brought with him the first haying equipment known to exist in the Methow—a horse-drawn mower and hay rake.

Thurlow and his son settled on the property along Beaver Creek that Mason had bought from Joe White on his first trip into the Methow the preceding spring. Thurlow's purchase for $22.50 was most likely the valley's first real estate transaction. Mason Thurlow became one of the valley's principal figures, admired for his leadership and community spirit as well as for his agricultural accomplishments. His descendants still live on the Beaver Creek property.

The Prewitts and Peter Filer also claimed places on Beaver Creek. Manford Stone and his daughter, Bertie, settled on Bear Creek near John Hartle's homestead. In the words of Bertie Stone, the site was favorable "because of the sparkling clear, pure water in the little stream and the abundant rich soil which could not be otherwise than productive."

Harvey Nickell, the fourth member of the Wise County exploratory group from the preceding year, returned with his wife, Alcena, and four children on July 4, 1888.

Alcena Nickell was probably the first white woman to come into the Methow. Her daughter, Mary Ellen, was the first white baby born there. Nickell claimed land for his homestead on both sides of Beaver Creek. He later donated parts of his claim to the community for the Beaver Creek School and the Beaver Creek Cemetery. Nickell's family, including his father and three brothers, played a significant part in developing the Methow's agricultural and commercial base in the decades that followed.

Walter Frisbee, another 1888 arrival, was the first person to claim water from the Methow River. A twenty-six-year-old bachelor, Frisbee developed the Frisbee Irrigation Ditch, later renamed the Foghorn Ditch, which originated near today's national fish hatchery. He built a ferry to cross the Methow River, but the spring flood of 1889 swept it away. Inventive and ambitious, Frisbee followed up with a self-propelled cable car for river crossings. The cable, which was anchored to the river's bank at each end, supported a passenger cage that ran on trolleys. The passenger would climb inside the cage and essentially propel himself across the river by pulling on a rope attached to the cage and to the far side. Frisbee established a variety of other businesses during the 1890s in the area of present-day Winthrop. Before long, however, wanderlust overcame him. He left the valley in about 1900 to explore and photograph the British Columbia interior. He came back a couple of years later, but in 1903 he sold his businesses and his Methow homestead to return to Canada.

A few of the 1888 Methow settlers claimed land on or adjoining what later became Moccasin Lake Ranch. Among this group were Sam Patterson, Dr. Granderson German and his son Will, Harvey and Ed Huss, and the Rader brothers, George and Pleas. (The Raders originally settled in the Carlton area in 1888 but moved to the Moccasin Lake Ranch area a few years later.)

As settlers moved into the Methow near the end of the nineteenth century, they faced an administrative hurdle that temporarily prevented them from filing a homestead claim. Because the Methow had previously been designated Indian territory, its lands had never been formally surveyed, meaning that there were no legal descriptions identifying the individual townships, sections, or individual parcels of land. Accordingly, settlers had no way to identify their own homesteaded quarter section. All that settlers could do was stake the corners of the land they wanted, hope that their claim comprised 160 acres, and then "squat" on the land until the government's survey team completed its work. This situation was remedied by the late 1890s, when land surveys in the valley had been completed, legal descriptions became available, and formal application for individual homesteads could be made. Surprisingly, there were few reports of boundary disputes, perhaps an indication of the spirit of interdependence and neighborly cooperation of the time.

Steamboat *Chelan* on the Columbia River at the dock in Wenatchee, Washington. Shafer Museum Collection

Chapter 5

Unbreakable

"United we stand. Divided we fall." This sentiment had meaning to Methow settlers, who banded together in difficult times. Neighbors helped each other, a trait that would continue into the new millennium.

The pioneer families that arrived in the Methow Valley in 1887 and 1888 were hardened travelers. Even before setting out from Ellensburg toward the Methow, they had already survived a bone-jarring wagon trip across the country from Texas or other distant states. Along the way they had encountered endless prairies, towering mountains, great rivers, and giant forests. Summer travel across the Great Plains in canvas-covered wagons could be unbearably hot and dusty. Great clouds of insects could appear out of nowhere, and occasional cloudbursts and electrical storms panicked the children and scattered the livestock. Travelers also faced the threat of rustlers or hostile Indians. By the time these settlers reached the Northwest, they were accustomed to overcoming obstacles, making do with very little, and surviving to tell about it.

The trip from Ellensburg to the Methow, though much shorter than the one that brought pioneers across the plains, was difficult, setting the stage for more hardships to come. The journey took about ten days using packhorses. With families, their possessions, and wagons, the trip stretched to more than two weeks. (Today we drive the route in less than four hours.) While crude ferries were used to cross the Columbia River, pioneers had to float across the Okanogan River in their own wagons. To prepare for this, wagons were soaked in water for more than a day to swell the wooden planks and thus seal cracks between them.

The Methow Valley would have been much more accessible had there been a river-level route starting at the confluence of the Methow and Columbia Rivers. But steep cliffs extended to the edge of the Methow River, cutting off access to the valley from its mouth at Pateros and forcing travelers to use the more dangerous routes across the Okanogan highlands, with its steep hills and deep ravines. The trip over the highlands, the area between the Okanogan and Methow Rivers, required almost seven days of difficult travel.

Three routes over the highlands were developed by settlers entering the valley before the turn of the century. The first wound through the Chiliwist Valley on the west side of the Okanogan River. This trail climbed up the east side of the highlands to a point near Loup Loup Pass. It then followed Benson Creek from its headwaters to its confluence with the Methow River about four miles south of Twisp. By 1890 a second and somewhat improved route had opened up. It was known as the Bald Knob Trail. This route left the Columbia River midway between Pateros and Brewster, where the Central Ferry had just been installed. It went up Indian Dan Canyon, past Bald Knob Mountain, and then connected with the headwaters of Texas Creek on the west side of the highlands. Settlers then followed Texas Creek to its confluence with the Methow River at Carlton. In 1891 a third route was developed. It started on the Columbia River at Brewster and went up Jasper Canyon past Paradise Hill. This route, known as the Brewster Mountain Road or the Paradise Hill Trail, also intersected Benson Creek in the vicinity of Loup Loup Pass and then followed the creek to the Methow River.

At the most dangerous points, travelers found themselves traversing narrow cliffside trails with grades as steep as 50 percent. Some ascents were so steep that passengers removed their belongings from their wagon and carried the goods up the hillside on their backs. On some descents, the wagon's wheels and running gear were removed and its box frame was skidded on the ground. In extreme cases, pine trees were felled and hitched to the back of the wagon to act as brakes. At steep ravines, trees dropped and positioned across the chasms served as makeshift bridges. The wagon boxes, with wheels removed, were then skidded across the ravine on the logs.

Once the pioneers reached the Methow Valley, they were geographically separated from the outside world. The nearest trading post was far away in the Okanogan Valley. Their safety and ability to survive depended primarily on their own fortitude, hard work, and ingenuity, yet each family knew that when help was required, a

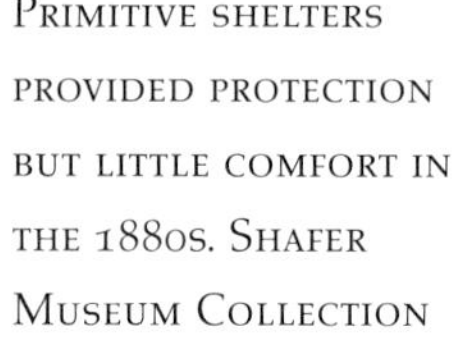

Primitive shelters provided protection but little comfort in the 1880s. Shafer Museum Collection

neighbor was available to provide it. The first priority was constructing a simple shelter. Early shelters were no more than rough log cabins with walls of pine, cottonwood, or aspen. Spaces between the logs were caulked with mud or clay. Small poles spanned from the top of the walls to a ridgepole. These were laid side by side to support the mixture of mud and clay that was plastered over the poles. On top, a layer of matted ryegrass or bulrushes was fashioned to keep out rain and snow. Cobblestones, which were readily found in nearby fields and streams, were used to build the fireplace and chimney.

Most of these early cabins had no floors. Hides from deer or oxen were commonly placed on the ground, hair side up, to provide insulation from the dampness. To facilitate sweeping out their cabins with crude sagebrush brooms, the settlers were careful to place these hides with the hair pointing toward the entrance.

Out of necessity, pioneers became skilled at using a variety of tools. These included the ax froe (a cleaving tool with handle at right angles to the blade), the broadax (a wide-bladed ax for making boards), and the drawknife. With these commonly used tools in hand, floorboards, roofing shingles, tabletops, and log siding could all be hewn from the nearby forests.

The first pioneers, with all the land in the valley at their disposal, chose their homestead sites wisely. They quickly selected sites on the Methow River's many tributaries rather than on the river itself, in order to avoid the river's dangerous spring floods, the dense willow thickets that hugged its banks, and the mosquitoes that infested these thickets. Although it may not have been apparent at the time because the Methow was going through a period of unusually high precipitation, the land claims adjacent to a reliable source of water would prove to be of great value in future years.

Creek water had many uses beyond cooking and drinking. The settlers built cold-storage sheds over their streams, using the chilling effect of the flowing water to cool perishables stored inside. On Saturday nights, great tubs were filled with creek water so that families, and their clothes, could receive a weekly scrubbing. In wintertime, stream water was ponded and frozen so that slabs of ice could be harvested for use later in the year. As settlement progressed, people began using creek water to irrigate their gardens and orchards.

Areas that contained stands of trees were also important. Trees were the raw material for fashioning tools and shelters. Trees gave protection to the livestock during winter storms and provided them with shade during the summer. Willow, cottonwood, and aspen were the predominant species along the river and wetlands, while stands of pine and fir were readily available on nearby hillsides.

From the earliest days, transportation represented a major challenge for Methow pioneers. Crossing the river was difficult except in late summer, a time of low water. Rowboats, canoes, hand-operated trams, and rudimentary footbridges all came into use during the rest of the year. But the trams and bridges generally didn't survive the spring floods. Some enterprising owners of footbridges tried to collect a small fee for their use, although cash-starved settlers usually forded the river on horseback or crossed in a canoe or rowboat to avoid the toll.

Trails and wagon roads were essential for keeping the settlers in touch with each other and with the outside world. No family could afford to be totally isolated from neighbors, whose assistance might be critical in times of emergency. Also, it was customary for neighbors to share in farm chores at certain times of the year. During the harvest season, men joined forces for haying. Crews moved from one settler's fields to the next as their crops came into bloom. The butchering of cattle and hogs and the shearing of sheep in the fall were often community affairs. Neighbors helped one another raise ridgepoles for their shelters and barns. Women served as midwives and nurses, providing medical assistance as best they could. All of these activities required travel, which focused the settlers' attention on improving roads and trails.

An improved wagon-road system also was needed to distribute the food, clothing, and equipment that were in growing demand. Only the crudest of wagon trails were available for transporting these items into the valley from Wenatchee or Ellensburg, the two closest towns. Settlers therefore spent much of the summers of 1890 and 1891 grading and smoothing the Bald Knob route. The next summer they worked on the Paradise Hill route. These citizen-financed undertakings were shining examples

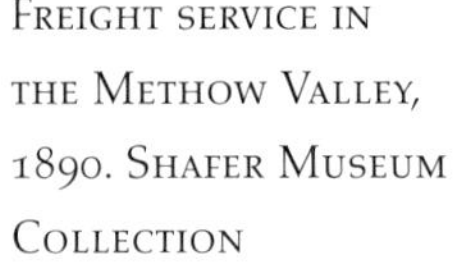
Freight service in the Methow Valley, 1890. Shafer Museum Collection

of how projects too big for any one person or family could be accommodated by neighbors banding together for the common good. This philosophy started early in the Methow and has remained a part of the valley's culture ever since.

Transportation outside the valley was also important to Methow-bound travelers. While the roads into the Methow from the Columbia River were being improved, the Great Northern Railroad was extending its line to Wenatchee from Spokane. Concurrently, the paddle-wheel steam vessel *City of Ellensburg* joined several other steamships on the Columbia River and began service between Wenatchee and Brewster.

These developments meant the journey into the valley was much easier than it had been a year or two earlier. No longer was it necessary to negotiate the 5,300-foot-high Colockum Pass, which separates Wenatchee from Ellensburg, and then travel through the Big Bend country by wagon train to the Central Ferry at Brewster. By 1893 travelers could take the Great Northern Railroad to Wenatchee and then board a steamer for an overnight trip up the Columbia to Pateros, then known as Ives Landing. There they would meet a stage that carried travelers and freight into the Methow over either of the Okanogan highlands' improved routes. The round trip from Wenatchee to Carlton was cut in half from almost a month to just two weeks.

These advances in transportation through the Columbia Basin brought many other changes. Settlers could now begin importing major items for their homes and businesses. Farm productivity grew as plows, harrows, discs, mowers, and hay rakes arrived. Sawmill equipment also arrived, enabling local mills to cut lumber for the settlers' cabins. Meanwhile, the production of ore concentrate at the valley's hard-rock mines, from Squaw Creek to Slate Creek, was approaching its zenith. These mines could now receive the heavy equipment they needed to improve production. The pace of life in the valley was gathering momentum.

By the early 1890s the area's roads had been developed enough to permit the Methow's first stagecoach and freight lines to operate between Brewster and Silver. The first of these lines, run by James Clark, was soon followed by another, operated by E. R. Davis. The ever-energetic Walter Frisbee, with his brother-in-law Cromwell Maxey, opened a third freight line in 1893 between Winthrop and Ellensburg. A fourth outfit was operated by Newt Williams. Teams of eight or more horses transported heavy equipment over the Bald Knob Trail, with spare teams ready at livery stables along the way. Peter Filer built a stage station near the summit of Brewster Mountain, where freight operators and their passengers could eat and rest along the steep Paradise Hill route.

Mail service to the Methow got under way in 1891 after Peter Filer contracted with settlers to bring mail once a week from the Okanogan mining boomtown of Ruby to the new town of Silver, its Methow counterpart. He was paid twenty-five dollars a month for delivering the mail, on snowshoes or horseback, to Silver's postmaster James Byrnes. The following year U. E. Fries, an immigrant from Denmark, subcontracted to carry the mail for $6.60 a week over an expanded route that included Malott in the Okanogan Valley and Winthrop in the Methow. Unable to negotiate a satisfactory room-and-board arrangement at Jimmy Sullivan's Winthrop hotel, Fries is said to have ended up sleeping on the ground a mile or so away to avoid Sullivan's hotel charges.

With mail and freight service and better roads, the Methow was becoming a less difficult and threatening place. Its residents were beginning to make the transition from survival to settlement. Yet life in the Methow was hardly luxurious — and a far cry from life in the big cities. While the valley's settlers were pleased with improvements to their log cabins, people in various metropolitan centers were enjoying such engineering marvels as the Eiffel Tower, the Brooklyn Bridge, the Washington Monument, and the Metropolitan Opera House in New York, all completed by 1890. While Methow settlers still lit their cabins with laboriously prepared candles made of melted tallow and cotton rags, some of their urban counterparts had electric lights and others enjoyed the use of gas streetlamps. Personal communication within the Methow could take place only through face-to-face meetings or the mail, far slower and less convenient than the telephones and telegraph available in eastern cities during those times.

However, these seeming disadvantages of the Methow did nothing to discourage the pioneers. In fact, many reported that the rigors of frontier life agreed with them. George (Ed) Nickell and his wife, Sally, came to the valley in 1888, where they lived in a small dirt-floor cabin on Beaver Creek alongside the cabins of three other Nickell men: Harvey, Wash, and Bud. One of Sally's granddaughters, Olive Mae Dibble, later learned from her grandmother that "despite the lack of luxuries of any kind, that first winter was the most enjoyable in her memory. They had plenty of food and wood to keep them healthy and warm, and they were young and very much in love. . . ."

Frontier life toughened the settlers and prepared them to cope with adversity as it arose. Perhaps the most important lesson taught was the benefit of banding together in difficult situations. Again and again during those early years, the settlers demonstrated that they could accomplish much by working together, both in everyday affairs and in times of trouble.

The settlers faced their first valley-wide crisis during the winter of 1889–90. The preceding two winters had been mild, with little snow and an early spring. Farmers were lulled into believing that their horses and livestock could continue foraging for feed in the hills throughout the winter until springtime, when fresh bunchgrass would be available. But after a modest snowfall in mid-December, winter set in with a vengeance. In early January, the thermometer dropped well below zero. A howling blizzard lasting almost two weeks swept through Okanogan County. Snowdrifts up to eight feet deep stayed on the ground until April. Intermittent thaws accompanied by rain showers and subsequent freezes formed hard crusts on the surface, preventing horses and cows from reaching the life-sustaining grass lying beneath the snow. Even the bunchgrass that was there had been stunted by the scant rainfall of the previous summer.

Families, caught with inadequate feed for their livestock, improvised mightily to keep their animals alive. They scraped snow from south-facing slopes and spread dirt on the exposed earth to promote thawing. They used penknives to cut grass and brush for the livestock. They melted ice in pans or chopped holes in ice-covered ponds to provide water for drinking. They even took the straw out of their own bedding and fed it to the starving animals. The horses nibbled each other's manes and tails in a desperate attempt to survive. George Rader is reported to have lost 23 of his 40 head of cattle but was able to save all of his horses. In the end, over half of the 150 cows and many of the 100 horses in the valley perished—and with them, a substantial portion of the valley's milk, beef, and transportation.

Food for the pioneers themselves also began to run low as the huge snowfall made travel in and out of the valley virtually impossible. "We had little to eat," Bertie Stone Filer wrote, "and before spring our supplies ran so short that starvation stared us in the face, and for over a month we lived on sour dough bread and venison. The deer were plentiful but the meat becomes almost unendurable when the diet is continued for any great length of time."

Neighbors rallied to meet the emergency, sharing whatever they could. Laura Thompson (no relation to George Thompson), who lived in a 12-by-16-foot cabin on Wolf Creek with her husband, Fred, and their son Roy, had received several sacks of flour before the storms set in. She reported: "All the settlers were out of flour by March, so each day some of our friends came to our cabin to see how we were fixed to lend flour. . . . Some got enough to run them until June. Bachelors took only one or two sacks." In spite of the close calls, discomfort, and loss of livestock, the pioneers kept up their spirits, passing the time during the storms by telling stories, reading books, playing checkers, and singing songs.

Less than a year later, settlers in the Methow were drawn into a difficult situation of a different sort. In early January 1891 word spread through the valley that an Indian boy had been lynched by a mob of miners in the Okanogan settlement of Conconully and that tribal members were preparing to retaliate. It was said that five hundred braves from the Colville Reservation were assembling in the Chiliwist Valley, ready to enter the Methow and massacre its inhabitants.

The fifteen-year-old boy, identified only as Stephen, had been taken into custody as a suspect in the killing of Sam Cole, a freighter who had been carrying goods to a trading post in Alma. Cole's badly decomposed body had been found in late December, inflaming passions and leading to warrants for the arrest of two young Indians, Stephen and a man named Johnny, who was in his early twenties. Johnny was shot dead in a gunfight at a nearby village. Stephen was taken to the Conconully jail until bail could be arranged. But in the wee hours of that night, about twenty masked men rode to the jail to take matters into their own hands. They tied up the jailer before leading Stephen away and hanging him from a nearby tree. This atrocity not only aroused the white settlers but also infuriated tribal members. Fears were magnified, and imaginations, in some instances, got the better of sound judgment.

The deadly defeat of General George Custer in 1876 by Sioux Indians at the Battle of Little Big Horn was still on people's minds, even though it had occurred some fifteen years earlier. Prompted by unsettling rumors about the warriors' intentions, Methow settlers began to mobilize. They started by building a stockade around Lee Fulton's cabin that they would use to protect the children and women. The men stayed up all one night, molding bullets and cleaning their guns. Women prepared for a long stay at the stockade by gathering needed belongings in burlap sacks.

On January 17, Lieutenant Governor Charles Laughton of Washington sent two hundred guns and ammunition to Conconully to counter the anticipated attack, and he ordered troops from Spokane Falls to be held in readiness. On January 19, General A. P. Curry of the state militia met with several Indian chiefs and warned them that his troops would make short work of them if necessary. He ordered them to stop their war dances, smoke signals, and the carrying of guns except for use in hunting. The chiefs assured the general that they had no intention of launching attacks. It seemed as if emotions on both sides had gotten out of hand.

The Indians returned to the Colville Reservation; the settlers returned to their normal lives. The "Indian scare" of 1891 passed without further incident. It did, however, serve to strengthen the bond among settlers, who again were reminded that their collective fate depended on working together.

The settlers confronted yet another crisis in the spring of 1894, when the Methow and the entire Columbia Basin experienced the worst flooding in memory. Drenching rains and warming weather combined to melt an unprecedented snowpack in the mountains and trigger an immense flood. Peter Filer's sister, Flora Filer, was returning to the Methow from Ellensburg when her trip was interrupted by flooding on the Columbia River. She wrote: "We were held up by high water at old Central Ferry, above Pateros. All ferries had washed out and we were camped there a week. Others gathered there and waited. Supplies were getting low and something had to be done. They got word to Wenatchee and the steamer 'City of Ellensburg' came and ferried us over. We watched the debris being washed down the river, logs, shacks and once a haystack with a rooster on top."

The flood killed one person in the Methow Valley. Frances Prebylowitz, sister-in-law of James Byrnes, fell into the raging current near the town of Silver when the bank gave way as she attempted to draw water from the river. The little town of Silver itself literally disappeared overnight, its few buildings washed downstream.

The Methow River waters also washed away the valley's first bridge wide enough to accommodate a wagon and team. The bridge, just downstream from Winthrop, had been built through a cooperative effort by the settlers and Okanogan County since there was no such thing as a state highway department. The settlers supplied timbers for the bridge's stringers, braces, and trusses as well as for the piers that supported the two ends. Settlers also provided volunteer labor, under the supervision of David Herstine, who lived on property that later became part of Moccasin Lake Ranch. Okanogan County supplied the decking and iron components.

Citizens were justifiably proud of their new bridge — a pride that turned to dismay when, just twenty-one days after it was completed, the bridge fell into the river and was swept away. In fact, all of the Methow's cross-river footbridges and conveyances were destroyed. However, settlers rallied together, salvaging whatever timbers, lumber, or other useful items remained so that structures lost in the flood could be replaced.

Sod-busting in the Methow Valley. Dick Webb collection

Chapter 6

Rhythms of Life

Farm tasks were numerous and demanding, requiring all family members to contribute in accordance with their abilities. Sundays were the times when adults relaxed and children played.

The lure of the West and the unrelenting promotions by the railroads continued to attract families to the western United States. Washington Territory was home to 75,000 people in 1880. A decade later, the territory had become a state and its population had grown to 375,000. By 1890, four years after the Methow Valley was opened for settlement, its population stood at 150, with most of these early settlers living between Carlton and Winthrop. (The entire white population of Okanogan County was approximately 1,500.) The Homestead Act, together with the completion of the transcontinental railroads, was accomplishing its intended purpose of simulating development of the West.

In the Methow Valley, family farming on 160-acre homesteads was replacing mining as the dominant activity. Many early arrivals consisted of families seeking escape from droughts in the Midwest and Southwest. Ras Garrett wrote that he was "looking for country where I can grow tomatoes and raise cattle on the same ranch." Other early arrivals were bachelors who simply wanted to stake a claim and start a family. Given a lack of mobility within the valley and the demands of improving undeveloped land, it was not unusual for a pioneer bachelor to become attracted to "the girl next door" and, in due course, take her for his wife.

The farms were labor intensive. Large families were the norm, and children were soon taught to help with the chores. The George Rader family had fifteen children — perhaps tops in the valley — but several others had eight or more. Farmers were generally not looking to expand their acreage; they had their hands full working the 160 acres of land they already held. Pioneer accounts emphasize how much manual labor was involved in developing a farm, and how the rhythms of family life were dictated by the seasons and by the everyday demands of frontier survival.

Certain jobs had to be done every day: milking the cows, feeding the livestock, cleaning the barn and cabin, hauling water from the spring, toting firewood from the shed. Cooking required considerable effort, but it was not unusual for girls of fourteen or fifteen to take over these household duties. By that age they had learned how to cook and bake and to can and preserve fruits and vegetables. Many young pioneer women also handled the weekly washing and mending of the family's clothes. It's not surprising that these young women often married before they were eighteen. Bachelors saw the obvious value of a wife who could run the household.

Most outdoor farm work came to a halt during the winter. In severe weather a family might lose contact with neighbors for weeks. During the winter of 1889–90, Laura Thompson did not see another person from Christmas Day until February 10. Lee Fulton wrote that after January 28 the snow was at least two feet deep and "the only means of conveyance was on foot." Given the distances between homesteads and the fact that most settlers were not accustomed to travel on skis or snowshoes, it wasn't unusual for families to simply stay in or near their small cabins.

Wintertime had its virtues, however. There was sledding over the crusted snow, and blazing fires that heated the cabin and dried wool clothing. Pearl Rader wrote that "winter was an easy time of year. We'd cut ice blocks from the frozen lake

Ice was gathered in winter to preserve food during warm summer months. Dick Webb collection

[for the food/storage icehouse], but that was more fun than work. We had enough wood for our stove and enough food to eat and father would sit and read us stories of the north, of Alaska."

As winter drew to a close and snow melted from the fields, farms became busy again. Preparation of virgin fields for planting was especially strenuous. First came land clearing, with trees felled by saw and ax, and stumps pulled out by horses or oxen. Then rocks in the fields, legacies of the ancient glaciers that had deposited them, were removed to prevent damage to plows and other equipment. Plowing, or sod busting, was done with a single plow pulled by a horse or oxen and steered by the farmer. Plowing turned over the topsoil and killed the weeds by cutting their stems. The operations known as disking and harrowing then broke up the clumps

The daily chore of feeding livestock. Shafer Museum Collection

of dirt produced by plowing and removed remnants of grasses that clung to the soil. All of these activities were completed while the ground was saturated with winter's moisture and thus easier to work.

The field was now ready for planting, originally done by hand. The farmer would grasp a handful of seed from a burlap bag secured to his waist and spread it in a semicircular arc as he walked across the field. An additional round of harrowing covered the seeds with dirt.

When the settlers first arrived in the Methow, they brought with them seeds for their own gardens and orchards. The abundant rainfall that blessed those early years helped produce hardy crops of potatoes, corn, rhubarb, and a variety of other vegetables — even without the benefit of irrigation. Orchards also did well. Apple and cherry trees were planted first. Later came plum, pear, and peach trees. Lilac bushes were favorites among many pioneers who enjoyed their fragrance and color. These bushes can still be seen standing alongside abandoned homestead cabins or places where cabins once stood.

Devising a system to bring water to gardens and orchards was an important part of pioneer work. Settlers diverted water from nearby creeks into hand-dug ditches. These ditches in turn irrigated the vegetable gardens and orchards. As irrigation was introduced around the turn of the century, orchards were expanded and additional fruits were introduced. While the men worked in their fields or on irrigation ditches, other family members planted or pruned fruit trees, worked in the vegetable garden, and watched over the springtime birthing of calves, foals, piglets, and lambs. No matter how large the family, the various jobs seemed to use up all the daylight hours.

As the weather warmed into summer, vegetables and fruits ripened. The vegetable garden and cherry trees yielded their crops first. Fruit was picked and either eaten fresh, canned in five-gallon crocks, or occasionally taken to town and sold. Vegetables were stored in pits lined with straw or pea fodder. Then alternating layers of vegetables (potatoes, cabbage, peas, carrots, and squash) and straw were laid down until the pit was filled. A mound of dirt went on top to preserve the pit's cool temperature but prevent the contents from freezing later in the year. The following winter, one of the children would go out to the pit and dig down through the snow and dirt to reach the layer that held the food that the family wanted. In this way, a family's diet could be varied throughout the year even in the absence of refrigerators, preservatives, and grocery stores.

The focal point of the farming year came with the annual summer harvest of alfalfa and grain. Alfalfa, a legume used to feed livestock during the winter, was cut and piled in haystacks. Early settlers cut the alfalfa by hand with scythes. The cuttings were then hand-raked into windrows that were pitchforked onto horse-drawn skids or platforms. By the mid-1890s, horse-drawn mowing machines capable of cutting five-foot swaths had been imported into the Methow — mechanical marvels that dramatically improved the productivity of each farmer. Increased crops allowed farms to feed more cattle through the winter, and herds were expanded accordingly. As hay production increased, so too did the size of the field crew, since a lot of hand labor was still required to gather and stack the hay as well as change the two-horse teams every few hours. Harvest work continued from daylight to sunset.

In the early 1900s the gathering of hay was labor intensive, but a necessary task for Methow Valley ranchers. Shafer Museum Collection

Early mechanical harvesting equipment. Shafer Museum Collection

Crews often used A-frame booms, about thirty-five feet long and attached to a mast by guy wires, for stacking the hay. The long boom, outfitted with a grapple and hoisting mechanism, could lift huge bundles of hay high into the air. The objective at that time was to build haystacks as tall as possible to minimize damage to the crop from summer rains.

Just before the turn of the century, wheat crops were introduced into the Methow. The grain harvest was also labor-intensive. A horse-drawn mowing apparatus cut off the stalks. These were then hand-bundled into sheaves and conveyed by wagon or sled to a central location for threshing. Threshing was performed on ground previously prepared with a thick covering of hay or straw. The sheaves were spread out and flailed so that kernels of grain were separated from their stalks and from the husks that protected them. Some early farmers ran semi-wild horses over the sheaves to thresh their grain. Others tethered teams of domesticated horses or oxen to a log-roller mechanism that did the same job. After separation, the kernels were carted away to a silo or granary for storage.

Sizable crews were needed to complete the harvest at each farm while its alfalfa or grain was in prime condition. Thus it was customary for farmers to trade off working on each other's farms, going from farm to farm to expedite the harvest. Visiting crews typically took their main meal at midday at the farm where a harvest was taking place. The women routinely prepared wholesome meals for crews that could number fifteen to twenty men. Children pitched in to help their elders in the fields, but they also knew how to have fun in the process. Lucille (Fulton) Peters had fond memories of summer harvest time, recalling "when we ran barefoot over the stubble to carry lemonade to the haying crew, the ride back on the hay-sled atop a huge mound of hay, the watermelon patch where I almost caught a pretty kitty that turned out to be a skunk that caught me, our hide-and-seek games with the neighbor kids amid the hundreds of corn shocks, the summer visits of our cousins from Twisp. . . ."

All too soon, summer turned into fall. The harvest was finished; the berries and apples had been picked and the potatoes dug. The leaves on the quaking aspens were turning golden. Children returned to school to start their sessions, which often lasted four or five months, until the following spring. It was time to make apple cider for winter and to cut and store wood for the fireplace and stove. In the late fall, fattened hogs and steers were butchered and deer hunted to supply hams, bacon, beef, and venison for the family's winter larder. In the absence of refrigeration, much of the beef and venison was cured in brine and then sliced into long, thin strips, dried in the sun, and smoked. These strips, known as jerky, provided winter meat and were popular because they did not spoil. After the pork was cured, hams and spareribs were cooked fresh, while bacon and shoulders were smoked and stored for winter. The fat sides were rendered into lard and generally kept in five-gallon cans for use as shortening. With these preparations complete, the farm families of the Methow were ready to settle in for the winter, contemplate the accomplishments of the past season, and begin their planning for the coming spring's work.

The Methow was a valley where farm families cared about their schools, churches, and civic activities. Settlers didn't intend to let poor roads, lack of school buildings, or limited teaching supplies deter them. They ensured that all their children received an education. The Methow's first schools were located in settlers' cabins. By 1890 home schools were operating at the cabins of Mason Thurlow and Thad Bamber on Beaver Creek, Lee Fulton on Bear Creek, and George Thompson and David Herstine on Thompson Creek. There were also home schools in the lower valley at the Williams, Metcalf, Rader, and Countryman cabins. Home-school cabins were primitive by most standards. They often had dirt floors, crude furniture, and only sparse teaching materials — perhaps an American flag, a dictionary, and some maps. Students wrote on slate found in the hills, using pens made of quill feathers from local birds.

To be a recognized school district, all that was required was a minimum of seven students. The student bodies generally consisted of the family's own children and perhaps those of one or two neighbors. In this regard, George Thompson's home school was unusual. Despite being a bachelor, Thompson opened his simple cabin to families so their children could be schooled there. Abbie Williams wrote that Thompson made a point of staying away from his cabin during school hours so he wouldn't disrupt classes.

Early Methow Valley schoolhouse. Shafer Museum Collection

The children routinely got themselves to and from school without help from their parents. This could require walking two or three miles, even in winter with temperatures below zero. Sometimes students had to cross the Methow River, often fording it on horseback or, during winter, riding a sleigh over the river's ice. Teachers at that time, some of them just a few years older than the students themselves, commonly moved in with a local family, paying room and board. A high degree of motivation was necessary to attend and learn in these early schools, but it is not surprising that Methow children overcame the various hardships and learned the traditional "three R's"—their numbers, their penmanship and, most important, how to read.

The pioneers placed such importance on schooling that several of them contributed their own land, lumber, or furniture to help get schools started. Robert Prewitt, for example, diverted lumber originally intended as flooring for his cabin to the Fairview School to complete its benches and desks. Harvey Nickell donated land along Beaver Creek for a school in that area, and Carl Johnson similarly donated some of his claim to the Winthrop school district. Virtually all the valley's earliest schools were built with 100 percent volunteer labor.

The Lakeview School, a log building erected near the present-day Moccasin Lake Ranch rodeo grounds, opened in 1891 with members of the Williams, Rader, and Herstine families in attendance. It was one of the first stand-alone schoolhouses in the valley. This school was soon followed by others at Beaver Creek, Rockview, French Creek, Libby Creek, and Texas Creek. By 1894, Methow citizens had completed ten schools. The settlers recognized that the school buildings could also serve as community gathering places. At Saturday night dances, neighbors shared potluck dinners as they socialized and frolicked. Amateur musicians played their mandolins, guitars, harmonicas, or fiddles to accompany the dancers who participated in the waltzes, quadrilles, schottisches, polkas, and minuets. Saturday night events also included literaries, gatherings where neighbors entertained each other with songs, musical compositions, speeches, or dances. Bertie Filer noted that people traveled as far as sixteen miles to participate, even when they had to ford the river on a cold, wintry night. On Sundays, valley residents assembled at one of the schools to listen to the impassioned sermons of a volunteer minister. Settlers were reminded by the Indian scare of 1891 that, in addition to being used for social or religious activities, schoolhouses could serve as stockades during crises.

Community events were usually well attended. In those days before there were telephones in the valley, the gatherings gave neighbors a chance to catch up on local news and discuss common concerns. Abbie Williams observed, "People were very friendly, and everyone was your neighbor from one end of the valley to the other."

Settlers also made time for civic and patriotic involvement. The Independence Day celebration of 1888 was probably the first valley-wide civic event in the Methow. Citizens gathered at a grove near Bear Creek, bringing horses bedecked with ribbons, wagons bearing flags, and copious quantities of food. Frank Fulton donated a three-year-old steer for the barbecue. George Thompson was selected — after some debate brought on by those who objected because of his Canadian birth — to read the Declaration of Independence and give the opening speech. The crowd loved it, and Thompson became the favorite speaker at many future public gatherings. After the speeches and a picnic, citizens settled into an afternoon of horse races, footraces, boxing contests, pie throwing, and the like. There was a prize for the best-looking young lady (won by Nell Stone) and one for the homeliest man (won by Mott Bryan).

In the fall of that year, the presidential election took place while pioneers were still in the midst of building their shelters and gathering supplies for the winter. Nonetheless, twenty-four men gathered at Mason Thurlow's cabin to vote in that election (eventually won by Benjamin Harrison, who defeated Grover Cleveland). Thurlow had bagged a deer with his Winchester rifle so the electorate could celebrate the occasion with fresh venison.

Two years later, after Washington had been admitted to the union as the country's forty-second state, citizen involvement was also very high in state electoral affairs. The most important issue on the statewide ballot that year was whether Olympia should remain the state's capital. Competitors for this distinction among Eastern Washington towns were Ellensburg and Yakima (the capital remained in Olympia). Sixty-five votes were cast in this election, again at Mason Thurlow's cabin.

The valley's residents were both active and energetic. They worked hard to improve their lot and that of the Methow as a whole. But they also needed some time off, and Sundays gave them that opportunity. After essential chores were done, horses were hitched to the hack or wagon and the entire family drove off to church. According to Vern LaMotte, a keen observer of the valley's development during almost the entire twentieth century, church services would be followed by huge picnic lunches brought by each family and consisting of "sandwiches, boiled eggs, potato salad, fried chicken, homemade ice cream, and usually three kinds of cakes, plus cookies." Afterward there were horseshoe contests for the men and baseball games for the children. Warm summer Sundays could also include berry picking, fishing, or hiking through wildflower-filled mountain meadows. One young settler remembered Sundays when "young people . . . would come to our place and boat on the lake. More than one romance blossomed."

THE 1910 CHRISTMAS BALL WAS HELD IN THE METHOW TRADING COMPANY HALL IN WINTHROP, WASHINGTON. SHAFER MUSEUM COLLECTION

Glover Street in Twisp,
Washington, circa 1900.
Dick Webb collection

Chapter 7

Taking Care of Business

The first settlements evolved into communities as businesses were established to service the valley's growing population and the Methow's wilderness retreated in the face of axes and plows of industrious farmers.

Supporting the Methow's growing population, commercial enterprises began to appear during the 1890s, as the valley's settlements expanded. Blacksmith shops were among the first. Smithies repaired wagons and wagon wheels that were banged up by the crude roads. They also repaired farm equipment and implements damaged by rocks in the fields. And they shod the horses that did so much of the work. One early smithy, Charlie "Blackie" Pennington, was so entranced by the opportunities in the Methow that he insisted that his two younger brothers, Frank and John, leave their home in Kentucky and join him in his Winthrop forge. They relocated to the Methow but ultimately went into the dairy business. Nels Peterson, another early Methow Valley blacksmith, had come to America from Denmark in 1877. His shop in Twisp was known to produce quality work, especially in the repair of mining gear and horse-drawn farming equipment. Nels, a man of enormous strength, continued to ply his trade until after his eightieth birthday. Then he retired to his home in Twisp, where he lived until his death a few years later.

Additional freight services also were needed, so in 1893, Pleas Rader joined the original teamsters — Clark, Davis, Maxey, and Williams — providing settlers with a choice of services to transport their goods between the Methow and various towns of the Columbia Plateau. By then, the Bald Knob and Paradise Hill routes over the Okanogan highlands had been gradually improved so that travelers and their goods could enter the Methow in relative comfort.

Other enterprises were opened in the 1890s to supply the valley's small but growing population. Sawmills began providing timber and lumber. According to Laura Thompson, Charlie Randall and John McKinney built one of the first sawmills, in about 1892, on Wolf Creek. Guy Waring and friends started a sawmill on the Methow River near Rockview a year or so later. Others appeared in the lower valley near Squaw Creek.

Hotels and inns sprang up about this time. Travelers and new arrivals could now use Jimmy Sullivan's hotel in Winthrop, the Burgars' Twisp Hotel, or Lee Ives' hotel at Pateros. Entrepreneurs eager to service travelers en route to the Slate Creek district opened hostelries at Ventura, Chancellor, and Robinson.

Cash was always in short supply in these frontier farm communities, so merchants had to be traders as well as salesmen. They extended credit, without interest, knowing they would receive payment when the farmer could afford it. Many merchants traded foodstuffs for fresh farm products or surplus farm tools. To get cash, some men left their farms in the summer to find work, although this required traveling to Ellensburg or the Big Bend country of the Columbia River. Wages from threshing grain in the Columbia Basin provided cash for foodstuffs such as flour, coffee, sugar, and lard, as well as seed and supplies for their farms.

Most of the goods produced in the valley were traded or bartered at the local trading posts. Exceptions were beef or hogs in excess of a family's needs. The carcasses of these animals were taken by wagon or sleigh to the Methow town of Silver, where they were transferred to a heavy-duty freight wagon for transport to Pateros. The carcasses were then loaded onto a paddle-wheel steamer en route to Wenatchee, and from there were sent on to Seattle by railroad. In due course, money would be sent back to the farmer after the meat was sold.

Everyone was self-sufficient in those days. Girls learned to make their own underclothing from flour sacks, and dresses from pieces of cloth. Clothing was handed down from older children to younger ones. Nothing went to waste. Yet it seems that no one felt unlucky or disadvantaged. Methow pioneers were too absorbed in their daily affairs to entertain thoughts of that kind. Pearl Rader remembered the day that

Livestock auction in Twisp, Washington, late 1800s. Shafer Museum Collection

she and her sisters "picked twenty-two gallons of gooseberries, currants, and dewberries and sold them to the town folks. The money we'd make from selling the berries in town would go for school clothes." Bertie Filer wrote, "Money was scarce as hen's teeth and often one pair of shoes and a gingham dress was about the extent of most girls' wardrobes and then we thought we were well-dressed."

In the midst of this frugal culture of farming and commerce, the Slate Creek mining boom and construction of Colonel Hart's remarkable wagon route brought a short-lived yet productive spike to the Methow's economy. This infusion of money from the outside couldn't have come at a more opportune time. It followed the devastating Methow River flood of 1894 and the depression caused by the repeal of the Sherman Silver Purchase Act in 1893. Mining activities gave the settlers not only much-needed income and improved transportation but also a great boost in morale during a time of need.

Before the end of the nineteenth century, a handful of settlements sprouted along the Methow River and became bases for commerce. Most of these communities are still with us today, including Mazama, Winthrop, Twisp, Carlton, Methow, and Pateros. Heckendorn survives only as a neighborhood on Winthrop's southern flank. Silver is long gone, a victim of both flood and economic distress, while the way stations en route to Slate Creek's mining district disappeared into history after the mines were abandoned. The following outlines the status of Methow communities in the late nineteenth and early twentieth centuries.

SILVER is generally regarded as the first Methow Valley settlement. It was situated between Beaver Creek and Benson Creek, near Polepick Mountain and the Red Shirt Mine. Silver ceased to exist in 1907. It maintains its stature in local history not only because it was the first Methow community, but also because it was the only Methow community to be completely destroyed by a natural calamity.

The first known white man in the Silver neighborhood was Jim Glover, a prospector who spent the winter of 1886–87 camped on the bench above the river. Glover was joined that same winter by fellow prospectors Chickamun Stone and Jack Forrester. The generally recognized founding year of Silver as a distinct settlement is 1890. Mine owner James Byrnes often gets credit as town founder.

From the time that Stone made his gold strike near Polepick Mountain in 1887 until the first years of the twentieth century, Silver played an active part in Methow history. Its businesses eventually included a general store, saloon with dance hall overhead, hotel, blacksmith shop, and post office. There was even a little one-room school. Silver was home to three or four local families plus all of the prospectors and mine workers of the area. Mail came weekly from the mining town of Ruby, located across the Okanogan highlands.

Then the Methow River flood of 1894 swept away Silver's general store, its hotel, and a few other buildings. Enough lumber was recovered to rebuild several structures, but this time the settlers built on a bench some twenty-five feet higher than the original town site. At the nearby Red Shirt Mine, where a new find showed promise of yielding ore containing both gold and silver, more than fifty men were recruited to work at wages of over two dollars per day. This bonanza lasted for more than a year, but the high cost of shipping and the chronic Methow problem of low-grade ore doomed the effort. As the mine's fortunes dwindled, so did those of Silver.

Still, the town was a freight-transfer point, had a post office where weekly mail deliveries attracted small gatherings, and was a natural meeting place midway between Carlton and Twisp. But the flood had relocated the river to the opposite side of the valley, almost a mile to the west. Thus the rebuilt Silver was literally high and dry at a time when competing settlements were emerging along the Methow River (Carlton and Methow) or at places where major tributaries entered the river (Twisp and Winthrop). Perhaps most important, the new wagon routes from the outside world entered the valley at Carlton and bypassed Silver.

By 1904 the nearby town of Twisp had received the post office formerly located at Silver. No longer on the river and without a post office, Silver was now a far less convenient meeting place. By 1904 only fifteen or twenty people still lived in the immediate area. The buildings of Silver were gradually abandoned. In 1907 the general store building, after standing empty for almost three years, was torn down and relocated to Carlton. Silver disappeared for a second and last time. The only reminder of its existence is a historical marker along state highway 153 that indicates where the town was rebuilt after the 1894 flood.

TWISP, at the confluence of the Methow and Twisp Rivers, was first officially known as Gloversville. In earlier times the site had been used for Indian encampments. H. C. Glover acquired land there around 1897. Glover went to the county seat at Conconully and platted the town site, naming it Gloversville. But in 1899, Amanda Burgar platted an adjacent parcel, giving it the name Twisp, after the river that flows into the town from the west. The townsfolk preferred Burgar's choice of names, and Twisp it has been ever since.

Much speculation has centered on the original meaning of the word Twisp. One opinion is that it originally came from Indian jargon for yellow jacket or wasp, since the word Twisp produces a buzzing sound. Another view has it that Twisp is the Chinook word for forks, as in a confluence of rivers. Still another notion is that it

Fourth of July celebration, Twisp, Washington, circa 1900. Dick Webb collection

means gathering place, since the area had long been a favored place for Methow Indians to camp. Regardless of the true origin, Amanda Burgar's decision to use the word Twisp trumped Glover's attempt to perpetuate his own name.

By the turn of the century, Twisp was the busiest of the Methow towns. It included Burgars' Twisp Hotel, a large blacksmith shop, a post office, numerous saloons, at least two trading posts, a drugstore, and several private homes. The town also hosted a fish hatchery, where thousands of salmon fry were reared and released to swim downstream to the ocean. Most of the business traffic was a product of the Red Shirt Mine's resurgence in the latter part of the 1890s and, to a lesser degree, the prospecting activity up the Twisp River. Miners had cash to buy supplies, and they bought them from Twisp's merchants, who carried a wider variety of items than the stores in other Methow settlements. By 1901, Twisp had opened the valley's first commercial bank—The Commercial Bank of Conconully—and also hosted the first doctor in the area. On August 2, 1909, Twisp residents voted 65 to 7 in favor of incorporation. The town has remained the Methow Valley's principal commercial center ever since.

Carlton, some eight miles downriver from Twisp, also grew throughout the 1890s. After settlers improved the Bald Knob route connecting the Methow with Pateros, A. C. (Linc) and Effie Fender opened a stage station along that important wagon road, which complemented Peter Filer's way station on Paradise Hill. It too provided a welcome place of rest and refreshment for early day travelers. As the Bald Knob route gained popularity after the 1894 Methow River flood decimated Silver and relocated the river, Carlton became the dominant settlement in the lower valley.

The village gradually grew to the extent that a town site was platted in 1907. It was named after a popular local resident, Carl Dillard, who was proprietor of the Locust Inn. Several well-known valley pioneer families selected homestead sites and settled in the area as early as 1888. These included the Rader brothers, George and Pleas, and the Risley and Libby families. Later the Williams, Countryman, and Reynolds families and others moved into the area.

The town of METHOW is located almost eleven miles downstream from Carlton on the Methow River. The town, which dates to 1894, was originally located on Squaw Creek, a tributary of the lower Methow River. For a time, the settlement itself was known as Squaw Creek. Then state authorities renamed it to avoid confusion with another town of the same name. The original community of Methow existed to serve the needs of workers at gold mines whose ore production had stimulated a boomlet in the lower valley during the 1890s. There were four stores, two hotels, a restaurant, a saloon, and a stable. During its peak mining period, there was even a sawmill and assay office. By 1897, however, mine output had begun to decline and the end of the town's existence on Squaw Creek was imminent.

In 1899, W. A. Bolinger relocated his store from the Squaw Creek site to a place on the Methow River. The store was now better situated to serve people traveling on the wagon trail along the river. Shortly after this move, the original town of Methow ceased to exist and the current town of Methow was born. It contained at least two stores other than Bolinger's, along with a restaurant and a post office. Soon a school was organized. In 1908, C. J. Ogden built a six-room hotel, which was followed by new irrigated orchards patterned after the one planted earlier by Jim Robinson. Today, the original schoolhouse, the stately Bolinger house, and the orchards are readily visible from state route 153 along the Methow River.

PATEROS is strategically located at the confluence of the Methow and Columbia Rivers. Originally Pateros was known as Ives Landing in recognition of Lee Ives, an early settler who farmed, raised cattle, and built a handsome three-story hotel on the town site. In 1900 the land surrounding the community, including most of the Ives holdings, was purchased by Charles Nosler. Seeing an abundance of waterfowl on the Columbia River directly in front of his land, Nosler renamed the settlement after a town in the Philippines that he had visited where thousands of ducks made their home.

Three to four hundred people lived in Pateros, and the community was developing into a transit hub for the Methow Valley. Arriving settlers had passed through the settlement for more than a decade. Most had taken one of the several steamships that had plied the upper Columbia River since 1888. Later, passengers could arrive in Pateros by the Great Northern Railroad. Pateros also handled all of the Methow's agricultural and mining products being shipped to Wenatchee and beyond. Cattle and sheep were driven to Pateros for loading onto the Columbia's stern-wheelers or railcars until the late 1920s. Mining concentrate was trucked there to waiting railcars through the 1940s.

Pateros served as the primary gateway to the Methow Valley from the early 1900s, when a river-level road was built up the valley, until completion of the North Cascades Highway in 1972. Even though the scenic northern route provides competition during summer months, Pateros offers year-round access to the Methow and still enjoys a healthy economy more than a century after its founding.

WINTHROP often dates its origin to 1891. It was then that Guy Waring returned to Okanogan County from the eastern United States and established the Methow Trading Company at the confluence of the Methow and Chewack Rivers. During the 1880s Waring had lived for several years in the Okanogan Valley near Loomis operating a cattle ranch, but he'd never crossed the highlands into the Methow Valley before 1891.

The first settlers were most likely those intrepid trappers the Robinson brothers, who built their cabin in 1886 near the present-day community baseball field across from Mack Lloyd Park. Jimmy Sullivan and his wife, Louisia Heckendorn Sullivan, followed the next year. The Robinsons soon moved farther up the valley, but the Sullivans stayed to build Winthrop's first hotel.

Walter Frisbee entered the Methow in 1888 and went on to operate a blacksmith shop, a trading post, and a photography shop, all where Winthrop now stands. Charles Look arrived early in 1891 and became the settlement's first postmaster. Mail was carried on horseback or by snowshoe to Look's cabin from the Okanogan Valley by U. E. Fries. Look's cabin served as the first post office in the Methow's upper valley.

After Guy Waring opened his Methow Trading Company store in Winthrop, he appealed to have the post office relocated from Look's cabin to his own store. It seems clear that Waring, an intelligent and ambitious man, was intent on also having the town named after himself, as early towns often took the name of the owner of the local post office. Waring most likely felt there was a good chance that with the post office in his store, the town would be called Waring. This was not to be, however, as the U.S. Post Office Department chose to name the post office for Theodore Winthrop. This man had written a book about his experiences exploring the Northwest some forty years earlier.

The department did, however, agree to move the post office to Waring's store. By 1897, Waring had exchanged a minority interest in his Methow Trading Company for the land contained in the town's plat. In 1901 the company started selling these sixty-nine lots, which form the heart of today's Winthrop.

The town of HECKENDORN emerged after Jimmy Sullivan's stepson, David (Gene) Heckendorn, was given a portion of the original Sullivan homestead. Neighbors urged him to plat the land as a competing town to Waring's neighboring Winthrop. County commissioners approved the plat of Heckendorn in 1904, and the fledgling town hosted Sullivan's hotel along with Chauncey McLean's general store.

Competition between the two settlements was fierce. The trading posts pitted Waring's Bostonian demeanor against McLean's folksy style. The local folks seemed to enjoy jawing with Chauncey and frequently patronized his establishment, which traded a variety of goods including fruits and vegetables sold on a commission basis. After years of competition and several disputes, in 1924 the communities of Heckendorn and Winthrop agreed to unite. Their combined populations were sufficient to meet the minimum size necessary to incorporate the communities as the town of Winthrop. Heckendorn disappeared as a town, although some residents still refer to the southern part of Winthrop as Heckendorn.

The town of MAZAMA lies beyond Winthrop, at the northern end of the Methow Valley. Before 1900 there were only a few scattered homesteads north of Winthrop. The first settlers were mostly prospectors working at the Slate Creek district mines or trappers like Jim and Tom Robinson. Hazard Ballard, Charles Boesel, and brothers Albert and Fred Ventzke moved into the Mazama area in 1889; the Wehmeyer brothers, Bill and Fred, and their families arrived in 1892 along with Henry Johnson and Charlie McClurken. Jack Stewart moved upvalley from Silver in the mid-1890s and George Vanderpool arrived in 1895. Only Boesel and Vanderpool farmed the land; the others mined, surveyed roads into the mine district, or operated sawmills.

Because of its minimal population, the upper valley wasn't awarded a post office until 1900. On June 1 of that year, Minnie Tingley was named postmistress. Until then the settlement had been known as Goat Creek, but the U.S. government changed the name at the time of its postal award. Some say that the word Mazama is Greek for goat, while others believe it may be the Aztec word for deer. Either interpretation is appropriate, as the area is home to both mountain goats and deer. In recent years the preponderance of the Methow Valley's new residential construction has taken place in the upper valley. Its convenient location at the eastern end of the North Cascades Highway and the beauty of its steep hills and forests are major attractions for valley newcomers.

There were other communities in the first decades of Methow Valley settlement. These included Rockview, Robinson, Ventura (which was only a tent camp), and Chancellor — all in the northern part of the Methow and all dependent on the mining activities of the Slate Creek district. Their colorful but short lives invariably revolved around the mines' fortunes and those of the characters employed at the mines. When the mining boom collapsed and the mines were shuttered, so too were the inns, eateries, and dance halls along the wagon trail to the mines. At the end of the century these communities gradually slid into obscurity.

At that time, the Methow Valley had completed its transformation from wilderness to settlement. Many amenities were still lacking, but settlers now enjoyed the services of tradesmen and a widening variety of supplies carried by merchants. It was possible to have a frame-built home constructed by carpenters and implements repaired by blacksmiths. Daily mail delivery was a reality. An ever-growing expanse of land was being cultivated. Orchards and gardens produced an increasing assortment of produce and fruit. A telling indication of the maturing of the Methow at that time was the creation of its first two cemeteries, at Beaver Creek and at Winthrop, a reflection of the aging of the earliest pioneers. By 1900 the second generation of settlers were at an age to stake their claims, raise families, and carry on in the footsteps of their forebears.

Winthrop and the town's new bridge during the flood of 1894. Shafer Museum Collection

Winthrop, Washington,
in 1904, a decade after the flood.
Shafer Museum Collection

Chapter 8

Growing Up

New roads connected the Methow to the outside world and brought many advantages to its citizens. Never again would the valley exist in isolation.

In the first quarter of the twentieth century, life in the Methow was abruptly impacted by events occurring in the outside world. It was a time of growth and change, brought on by better roads, bigger towns, mechanized agriculture, and introduction of the automobile, followed by the trauma of World War I.

Transportation was still a key issue. The rough topography combined with heavy winter snows continued to hinder travel. Runoff from melting snows, which rendered local wagon trails all but unusable in the spring, sometimes demolished even the sturdiest pioneer bridges. To the west, the heavily forested slopes of the harsh Cascade Range discouraged travel; to the east, the steep hillsides and deep ravines of the Okanogan highlands offered only the roughest of wagon roads. The appealing idea of a road that ran along the Methow River seemed an impossible dream. The eighteen miles of granite cliffs that extended to the margins of the river, starting at a point just above its confluence with the Columbia River, appeared impenetrable.

Methow residents felt the squeeze. Families did not have easy access to local merchants who stocked the goods and clothing they wanted. Sawmills producing lumber couldn't readily deliver their products during the winter months. Barbers, butchers, doctors, lawyers, and merchants all required more reliable transportation, and the school year remained shortened to minimize the times that children had to ford the river or travel on horseback over uncleared trails during winter.

A consensus developed that transportation was the number one problem in the valley. And while a solution was beyond the financial ability of the Methow citizenry, and even of Okanogan County, a man of great capacity emerged in the Methow. His name was Walter A. Bolinger.

W. A. Bolinger had settled on Squaw Creek in 1892 at the age of twenty-nine and opened a mercantile to supply miners. He soon recognized that the potential of the Methow Valley lay not in mining but in agriculture. This realization, along with the decline in mining, prompted him to move his store to a site a few miles away, on the Methow River. Next he secured the local post office at his store. This was important because citizens gathered whenever a mail delivery was scheduled. Bolinger, a shrewd man who harbored political ambitions, recognized the mail gatherings as opportunities to sell merchandise and to establish himself as a man who could help the valley as an elected representative in the state capital.

Bolinger ran for the state legislature in 1904 on a platform of better roads, and he won. During his three terms in Olympia, he successfully championed bills that allocated thousands of dollars to the little-known Methow Valley, including funds for a lower-valley road along the river. Experts questioned the feasibility of such a road, which would require blasting through miles of granite. But Bolinger doggedly pursued the project. By the time he left the legislature in 1916, seventy miles of unpaved road through the Methow Valley connected Pateros to Robinson Creek. Work on the road was carried out by an unusual conglomerate of private contractors, inmates from the state's Walla Walla penitentiary, and local volunteers. At first the route was only a rough, single-lane wagon road. However, it was improved as time went on and had been paved by 1938.

The new road had a great impact on the Methow's social and economic life. Neighbors could more easily visit one another or assemble for socials, literaries, or religious services. Machinery too heavy or cumbersome to transport over the steep pitches of the Okanogan highlands could safely be brought along the river road to sawmills and mines. Bridges built by the state's newly formed Department of Transportation permitted river crossings even during high waters. Transportation within the Methow, once limited to pack ponies and horseback along rudimentary trails, was transformed as stages, freight wagons, and carriages appeared.

Perhaps the most profound impact of the new river road was that, for the first time, the Methow Valley was conveniently connected to the outside world. Passengers and freight could travel from any part of the Methow to Wenatchee in less than two days. The trip to Pateros could be accomplished in a single day. No longer were valley residents cut off from events and developments in other regions.

The lives of local residents were further enhanced around the turn of the century by the delivery of news and mail every day of the year. The valley's first newspaper, the *Methow Valley News,* began publishing in 1903, followed by the short-lived *Winthrop Eagle* in 1905 and its successor the *Methow Valley Journal* in 1912. Telephone service connected Silver with the settlers on Beaver and Frazer Creeks beginning in 1902. Telephones were available valley-wide by 1910. The citizen-built system overcame the lack of telephone cable with typical Methow ingenuity. It used what was available — barbed-wire cattle fencing — to conduct the sound of the human voice.

In Twisp, a grand entertainment center known as the Opera House was built in 1903, complete with balconies and stage. The building became a popular place for

Originally constructed in the mining camp of Squaw Creek, this one-room schoolhouse was moved to the town of Methow in 1899. Photo by Barry Provorse

dances, band concerts, plays, minstrel shows, and eventually silent movies. Some of the Methow's earliest pioneers and leading citizens helped finance the showplace, including Joshua and Lon Risley, George (Ed) Nickell, P. L. Filer, and Jim Robinson. Robinson ended up with shares in the enterprise only because he discovered too late that the promoters didn't have the cash to pay for the lumber he supplied from his sawmill. When the Opera House burned to the ground in Twisp's great fire of 1924, Robinson's share became worthless and he commented that he would have been better off if he had never seen a sawmill. However, Twisp survived the fire and upheld its reputation as the most progressive town in the Methow, rebuilding its bank and adding a creamery, doctor's office, livery stable, and other establishments.

Downriver, the growing settlement of Carlton was serving lower-valley families with its post office, fourteen-room Locust Inn, a one-room school, and a blacksmith shop. Carlton merchants traded with the families living on nearby Texas, Gold, and Libby Creeks as well as with travelers entering the Methow via the Bald Knob route. In the spring and fall each year they also supplied lodging and food to the cowboys and herders taking their livestock to and from the Columbia Basin.

Winthrop continued to serve the needs of most of the upper valley. By the early 1900s Winthrop was a well-recognized trading center with its own newspaper, creamery, blacksmith shop, hotel, and saloon. In 1912 the town's citizens constructed a handsome brick schoolhouse to accommodate an expanding population of children.

Guy Waring, with his chain of five stores, was the most notable of the Methow's early merchants, but there were others. Dick McLean, Dan McAlister, George Nickell, and William Magee all supplied both hardware and groceries at their mercantile establishments in Twisp or Winthrop. Merchants were pleased with the valley's growing population and no less by the fact that residents were increasingly paying for their goods in cash. The valley's fledgling industrial and commercial establishments, by this time, paid cash to their employees, many of whom were also part-time farmers or farm workers. Their paychecks gave them badly needed supplementary income and alleviated the need to go outside the valley to find cash-paying jobs.

During the first half of the twentieth century, agriculture remained preeminent in the Methow economy. Industry, represented primarily by sawmills and sporadic mining activity, was essential but never as dominant as it had been prior to 1900. George Fender and Guy Waring had their sawmills at Rockview and near the Weeman Bridge before the turn of the century. Fender and C. N. Burton built a sawmill close to Moccasin Lake Ranch, in the Elbow Canyon area, around 1910. Other mills were constructed about the same time on the Twisp River and on Gold and Texas Creeks. These mills produced a wide variety of products — lumber for houses and barns, shook for apple boxes, furniture and other household items, as well as telephone poles and mine and bridge timbers.

Farm production reached new highs after steam-powered threshing machines appeared in the valley about 1903. These wood-fired, smoke-belching behemoths used steam to drive the threshing equipment. A steam engineer and a crew of eight or ten men were needed to operate the equipment and keep the boiler supplied with water and the fire box filled with wood.

Most of the grain produced in the Methow was ground into flour for domestic use. This led Harvey and H. L. Lewis to construct a roller mill in Twisp in the early 1920s to produce flour from locally grown grain. A waterwheel, turned by water carried from the Twisp River in an elevated flume, powered the operation. Unfortunately, its product did not compare well with flour produced by the mills in Wenatchee. After a short period the Lewises closed the mill, and Methow farmers returned to the practice of transporting their grain out of the valley for processing.

While the earliest settlers claimed land along the tributaries of the Methow River, later arrivals had difficulty finding unclaimed land adjacent to surface water. To address their need for water, they formed water cooperatives and began to engineer large-scale irrigation projects. Before electric pumps became available, these projects depended on gravity flow; hence only land below the elevation of the main ditch could benefit. Each of the farms in such a project would build a ditch through their land along a surveyed route. Work was carried out simultaneously on all properties so that each section of the entire project would be completed at the same time.

View of a more settled Methow Valley, 1920s. Shafer Museum Collection

Irrigation helped assure that crops could be raised on an annual basis regardless of rainfall. Irrigation also allowed farmers to diversify their crops. Between 1903 and 1911, the Skyline, Chewuch, Foghorn, and Fulton ditches were all put into operation by various groups of neighboring farmers, irrigating some ten thousand acres of what had been dry land. The Wolf Creek Reclamation District, supplier of water for almost half of Moccasin Lake Ranch's irrigation needs, was completed more than a decade later, in time to start the 1924 season.

Widespread irrigation allowed the valley's farmers and ranchers to grow more alfalfa and feed significantly larger herds of cattle. It was not uncommon for individual herds to grow to 100 to 150 head. As the size of herds grew, cattlemen began to drive their surplus livestock to Pateros for shipment by rail to Seattle. The drives generally took three days and occurred mainly between 1900 and 1930. After that time, cattle were transported out of the valley by truck on the new road along the river.

Raising dairy cattle was often limited to the number of milk cows necessary to supply a family's own needs for milk and cream. In this time before mechanized equipment, dairy cows had to be milked by hand twice each day. Careful not to let anything of value go to waste, families took surplus milk and cream to creameries in Winthrop or Twisp each Monday morning. There the surplus could be sold or exchanged for products such as eggs and butter.

Sheep were introduced to Okanogan County in the early twentieth century. Until then, cattlemen had enjoyed almost exclusive use of public lands for grazing their livestock. Exclusion of sheep had the additional advantage of minimizing the need for rangeland fencing. (Ownership of cattle could be determined by their brands.) However, after a modest count of about two thousand sheep in 1900, the number of sheep in Okanogan County exploded to almost thirty-two thousand by 1904. The sheep were viewed by cattlemen as trespassers on what until then had been their private domain.

Hostilities grew between cattlemen and sheep ranchers as each side erected fences to prevent grazing on land for which they claimed proprietary rights. Tensions boiled over periodically, as they did one winter night in 1903 near Malott, in the Okanogan Valley. A man named C. C. Curtis had allowed his sheep to graze on land owned by a widow who used it to pasture her milk cows. In spite of the woman's request that he stop this practice, Curtis continued to trespass. Finally a group of angry cattlemen killed some nine hundred of his sheep in what became known as the Curtis Sheep Slaughter. (A historical marker along state highway 20 in the vicinity of Malott indicates the site of the incident.)

Over time, passions subsided and bands of sheep became more commonplace. In the Methow Valley, Frank and Bertha Morse, owners of Moccasin Lake Ranch from 1925 to 1961, developed one of the area's largest sheep herds. In the 1920s the federal government concluded that good conservation practice mandated fewer animals in the forests. Grazing permits were issued that restricted the number of animals in an area. Later, large areas of the forests were closed altogether to grazing.

In the midst of these Methow Valley developments, nature stepped in with a humbling reversal. Until 1917 valley settlers had experienced several decades of above-average annual moisture. In these wet years, even the unirrigated uplands were lush with bunchgrass, and crops could be grown there. These circumstances lulled dryland settlers into believing they could successfully produce crops each year without having to irrigate or fertilize.

They were wrong. The year 1917 marked the beginning of seventeen years of severe drought. Wells and springs gradually dried up. The unfertilized soil, deprived of nutrients and water, yielded poor crops. Families could not make payments on their crop loans, forfeitures and bankruptcies abounded, and the banks and insurance companies reluctantly became major owners of upland properties.

Many of these farm families simply walked away from their homesteads to seek work elsewhere. Some left behind their farm equipment and horses. The onset of the drought spelled the beginning of their demise as farmers in the valley.

The Methow was becoming segregated, separating those with water from those without. Settlement in the valley became concentrated along a green beltway adjacent to the Methow River and its tributaries. The Methow remained a desirable place to live and to farm during this period for those families with access to water. But on the dry hillsides to the east and west, one could see only settlers' abandoned cabins and cattle and deer feeding on the sparse grasses and shrubs.

During this time, it was not uncommon to see roving bands of abandoned horses on the range. However, in a few seasons the horses consumed much of the bunchgrass that had made the valley so attractive to the original settlers. The absence of bunchgrass permitted bitterbrush to proliferate, a development that attracted thousands of whitetail and mule deer. In an ironic twist of nature, the valley's deer then became so numerous that by the 1940s the state's game department began to erect fences not only to protect the growing number of orchards in the Methow but also to allow the depleted stands of bunchgrass to reestablish themselves.

The Methow Valley saw its first automobile around 1910. According to Lee Fulton, that was also the year of its first auto accident. Fulton reported that a newly arrived Methodist minister lost control of his car near the top of a steep grade after the motor died and the brakes failed. The car plummeted down an embankment into some underbrush. Both the minister and his passenger escaped serious injury.

Sweat and black powder carved the first rugged road between Winthrop and Pateros, circa 1915. Dick Webb collection

During this same period, two doctors were both seeking the distinction of having the fastest conveyance to provide their services. Dr. Couche had a car that used a crank not only to start the motor but also for steering. Dr. Lawson purchased a motorcycle because he thought it would be faster than Couche's car. Neither vehicle ran reliably enough to warrant continued use. It was not long before patients awaiting house calls pleaded with the doctors to return to the more predictable horse-drawn carriage.

The Ford Model T, which cost $675 in 1913 and $625 the following year, was becoming increasingly popular. In 1914, George Nickell and his son Ben opened a store in Winthrop and were among the first using trucks to haul merchandise into the area via the new river road. Two years later, Leonard and Paul Therriault purchased three motorized vehicles — a Model T truck with hard rubber tires, a two-ton Velie truck, and an eleven-passenger Reo Speed Wagon — to carry the daily mail, freight, and passengers into the Methow. By the time of the First World War, the Methow roads and the reliability of the automobile had improved. There was no holding back general acceptance of modern transportation.

In the second decade of the twentieth century, the long tentacles of World War I wound their way from Europe into the distant Methow Valley. In June 1917, a Home Guard regiment was formed and some 175 young men from the Methow Valley were registered. Of these, forty-two went into active duty, including sons of some of the Methow's earliest pioneer families — Boesel, Dibble, Fender, Haase, Heckendorn, Hotchkiss, McLean, Nickell, Prewitt, Sullivan, and Williams.

Young men who left the Methow for the front in Europe were introduced to places and cultures they had only heard of or dreamed about. Fighting a war and seeing Europe were unique experiences for any farm boy from the Far West. Many of these men returned from the war with technical or professional training that they found useful in civilian life. Butler Woodward, for example, received training from the Navy in the distribution of electricity. After the war Woodward and his wife moved to Winthrop and established the Upper Methow Valley Power and Light Company. It was part of Arthur & Fowler Company, Electric Water Heaters, a modest conglomerate based in Spokane. Winthrop got its first electric power in 1923, thanks to Woodward, who oversaw the building of a dam, two spillways, a power plant, and several miles of transmission line. Twisp initially received direct-current electric power from storage batteries in 1911, but it was 1928 before standard alternating-current electric power became available to that town's citizens. Although electricity from these initial efforts was sporadic and often temperamental, residents lived with it until the federal Rural Electrification Administration came to the Methow in 1940 and bought Woodward out.

World War I produced its share of the technical innovations often associated with such cataclysmic events. Vern LaMotte praised the battery flashlight, a product of the war. He wrote that flashlights were a great improvement over candles for lighting nighttime visits to the outhouse: the wind couldn't blow them out. Vern also enthusiastically endorsed the advantages of toilet paper, another innovation of the times. He found it to be an improvement over the previously used pinecones.

BIG AUCTION

SALE

I will sell at my place in South Winthrop, on

Monday, April 28, 1930

Commencing at 1:00 o'clock p. m.

Dairy Cattle 11 head all Jerseys—1 Cow 8 years old, 1 Cow 7 years old, 2 Cows 6 years old, 3 Cows 4 years old, 1 Cow 3 years old, 3 Heifers 2 years old.

Horses, Hogs 1 Team 11 and 12 years old weigh 1400 pounds. 7 head of Hogs.

Implements Mower and Rake, Log Chains, Melotte Separator, Spring Tooth Harrow, Saddle, and numerous other articles.

TERMS—$10 and under cash; over that amount 6 months time on bankable note drawing 10 per cent, 5 per cent off for cash over $10.

Col. A. L. Fox, Auct.
L. D. Hollaway, Clerk.

C. J. Heckendorn

The varied terrain of Moccasin Lake Ranch included creeks, valleys, mountainsides, aspen groves, and grassland.
Photo by *Methow Valley News*

CHAPTER 9

A Ranch Is Born

As an oak grows from a tiny acorn, so too did Moccasin Lake Ranch enlarge through numerous expansions during four ownerships.

Until the early twentieth century, there was virtually no large consolidation of land in the Methow. Land ownership throughout the valley and around today's Moccasin Lake Ranch consisted of 160-acre homestead grants. Early settlers simply were not inclined to seek additional homestead allotments. The rule of thumb at the time was that a farmer could plow about two and a half acres each day with a good mule or team of horses. This physically taxing work filled the days from dawn to dusk. Thus, a large proportion of a farm family's summer was spent preparing, planting, and harvesting crops from a single homestead allotment. There was little time or energy left to contemplate taking on a second or third.

At the heart of what is now Moccasin Lake Ranch is the land settled by George Thompson in 1887. Originally the area was known simply as the Thompson place. A decade later, government surveyors assigned the name Moccasin Lake to the body of water adjacent to Thompson's claim. They apparently thought the lake was shaped like the Indian footwear. When T. D. Johnson acquired land in 1905 that included the lake, he appended its name to the original Thompson claim, which he had bought several years earlier. Since 1905 the property has been known as Moccasin Lake Ranch.

It is not known exactly where Thompson built his simple shelter. Extensive farming has obliterated any evidence of the cabin. Most likely he chose a site along the east side of the creek that now bears his name. That would be at the southeast corner of his homestead, in an area of north-facing hillocks where he would have had year-round access to creek water and to the open fields beyond. He would have been close to the timber along the margins of the nearby wetlands, a source of logs for his cabin and firewood for heating and cooking.

Abbie Williams, who attended school in 1890 at the age of eight in Thompson's little cabin, described it this way: "a one-room shack dug back in the hill and the floor was dirt." Digging into a foothill or hillock to make space for a shelter was a common practice of the earliest settlers. They used hills to shield their shelters from the cold winds that blew out of the Cascades.

George Thompson was a man of many talents: capable farmer and accomplished orator, square-dance caller and thoughtful neighbor. He opened his cabin for use as the area's first schoolhouse. During the severe winter of 1889–90, he tramped some six miles on snowshoes to deliver flour to the isolated Wolf Creek cabin of Fred and Laura Thompson (no relation to George). Thompson also was one of the pioneers who volunteered to help build the valley's first wagon road over the Chiliwist Pass in 1888.

Thompson was also an expert in the mountains. He is believed to have been the guide for Owen Wister during the author's 1901 hunt for mountain goats in the North Cascades. Wister, famous for his novel *The Virginian*, chronicled his Cascades adventure in the book *The White Goat and His Country*, in which he referred to his guide as "T"—an initial widely thought to stand for George Thompson. Wister told readers that "I have never hunted with a more careful and thorough man," describing how "T" always found the best routes over dangerous terrain.

Thompson, however, was mainly busy developing his 160-acre homestead and cultivating the land. It is not known whether he improvised any form of irrigation using water from the creek. If so, he may have raised alfalfa, which was introduced in the Methow in 1890.

Importantly for Thompson's successors, he did claim water in 1888 from the creek. This positioned his claim in front of those that came later, giving him Class A water rights. After Thompson had fulfilled the requirements of the Homestead Act, the U.S. government granted him a land patent (deed) to his claim on May 16, 1902. Six months later, he sold his homestead to T. D. and Mary Johnson and moved to the fledgling village of Twisp. There, Thompson bought four lots from Joshua Risley, one of Twisp's leading citizens. He lived in Twisp for most of his remaining days.

George Thompson was the only resident in the Moccasin Lake Ranch area during the winter of 1887–88. His nearest neighbors were Jimmy Sullivan and his wife, Louisia, who lived about four miles away in Heckendorn. During the next few years, however, Thompson welcomed a few new neighbors who each claimed a homestead. They included Sam Patterson, Harvey Huss, Granderson German, and H. H. Williams. Additional neighboring homesteads were later claimed by Pleas Rader and E. L. Boyd. All of these 160-acre homesteads, except for Sam Patterson's, eventually became part of Moccasin Lake Ranch.

The history of ownership before full consolidation of the ranch's lands is a complex one — a tangle of sales and purchases, settlers coming and going, names and dates and boundaries. The following summarizes the backgrounds of the earliest families in the vicinity of the ranch.

Patterson Lake in the 1920s. Shafer Museum Collection

Sam Patterson, born in Walla Walla, Washington Territory, in 1866, was a scout traveling with the wagon train that carried so many of the Methow's first settlers from Wise County, Texas, to Ellensburg in 1883. He arrived in the valley in 1888 as a bachelor, but three years later married John Hartle's oldest daughter, Elsie. Patterson had met the young girl on the trip from Texas. They settled on the south shore of the lake later named for Patterson (and below the mountain that also received his name). Abbie Williams, a neighbor, remembered Patterson as "tall and slender, always wore wrapped canvas leggings, a wide cowboy hat and had a brown mustache." He also was known for his fine sorrel Morgan stallion named Old George. The pair made an impressive sight around the valley.

Patterson staked a claim for water from Patterson Lake in 1888, the first such claim for water from the lake. He subsequently built an irrigation system for his alfalfa fields and orchard. In 1895, Patterson sold out to George Rader for $350 and moved his family to the east side of the Methow River, settling near Beaver Creek.

Harvey Huss and his son Ed also arrived in the Methow in 1888. The Huss land on Thompson Creek, about a half-mile east of today's rodeo grounds, extended eastward to Dibble Lake and included the alfalfa field southeast of Twin Lakes Road. This claim was sold to David Herstine, most likely in 1890. (Records of the transaction were lost in a courthouse fire in Waterville, Washington, where local records were kept at the time.) Herstine received his land patent (deed) and, in 1900, sold his property to J. C. Garrett. A portion of this property was eventually acquired from Jack Barron by Moccasin Lake Ranch.

Granderson German, a Civil War physician, was a widower who also traveled with the Wise County wagon train. He was an old man by the time he moved to the Methow in 1888. German and his son, Will, settled near Twin Lakes. In spite of his age, he was the doctor in attendance at many of the births to early Methow settlers. Twenty-four-year-old Will soon returned to Ellensburg to marry Pernina Houser and bring her to the Methow, where they lived with Will's father. Their claim included a productive alfalfa field on the opposite side of Twin Lakes Road from the Huss claim. It extended northward to the present-day site of Liberty Bell High School. After only four years in the Methow, the Germans returned to Ellensburg to take advantage of amenities and services available to the aging Dr. German. They sold their claim in 1892 to John and Rosa Ford. Shortly after the turn of the century this property also came under the control of the Garrett family, which incorporated it into their widely known One Bar Ranch. After passing through various ownerships, it was also acquired by Moccasin Lake Ranch from Jack Barron.

H. H. Williams, born in Pennsylvania, walked into the Methow Valley from Waterville, Washington, in the spring of 1889. That fall he sent for the rest of his family, which included his wife, Susan, their son Newt, and Newt's wife, Clara. They moved into his cabin on the south side of Patterson Mountain by the Elbow Coulee turnoff. This property included groves of tall cottonwoods and aspens, good open farmland, beaver ponds, and a reliable spring. Williams remained on the land until his death in 1918. His wife continued to live there until her home was destroyed by fire in 1923. The property was then sold to Eric Erickson, who built a new house and lived there until 1929, when he was killed in an auto accident. Erickson's heirs sold the property to Frank Morse, who incorporated it into Moccasin Lake Ranch.

E. L. Boyd homesteaded one of the last unclaimed tracts in the area shortly after the turn of the century. Boyd received his deed in 1909. His land lay immediately east of George Thompson's homestead. It extended from the present rodeo grounds on the north to an upper Wandling Mountain bench on the south. (It included the site of the current Pigott home at the ranch, as well as several ponds on the upper benches.) Boyd located his cabin on Thompson Creek, alongside the rodeo site, where he lived until selling the property in 1929 to Frank Morse.

The Rader brothers, George and Pleas, came to the Methow in 1888 and settled for a time in the lower valley near Carlton. The brothers had married sisters Hily Ann and Addie Barnhart. In 1895, George Rader bought Sam Patterson's homestead.

After a year of farming in the Columbia Basin at a wheat ranch near Davenport, Pleas Rader returned to the Methow and claimed land on George Thompson's southern boundary. The brothers and their families were reunited, this time near Moccasin Lake. A few years later, Pleas sold his property to T. D. Johnson, the man who bought George Thompson's homestead in 1902.

A story involving George Rader swept the valley shortly after he had purchased Sam Patterson's place in 1895. It seems that George had volunteered to work on a road crew improving the valley's wagon trails. One day the crew encountered the Mathias Walter family coming up the trail and noticed musical instruments, including an organ, in the Walters' wagon. Rader asked if the travelers would play the organ for the road crew if the crew unloaded the organ and placed it on clean boards

THE RADER HOMESTEAD AT PATTERSON LAKE. SHAFER MUSEUM COLLECTION

so that it wouldn't ingest dust from the road. The family agreed, began to play, and soon had the entire crew dancing like couples at a Saturday night dance. A few weeks later George spotted the newly arrived family at a Saturday social and quieted the crowd to make an announcement. He introduced the Walters, proclaimed their musical capabilities, and once again asked that they step up and play—this time with borrowed instruments. The Walters obliged and instantly became a great hit in the valley at dances and other festive occasions.

For many years the Rader brothers farmed and raised horses for freight hauling. As teamsters, they hauled equipment and supplies to the area's mines and settlements. The family is remembered in the valley for its part in ranching, mining, construction, and hauling freight for half a century.

Land consolidation in the valley began about 1900. Many of the original settlers had fulfilled their homestead ownership requirements and received their land patents. With legal title to their property, these settlers were in a far better position to buy or sell land than previously, when they had only staked a claim to it. Prospective buyers could feel secure purchasing deeded and developed land. Furthermore, using modern farm equipment, second-generation Methow residents could manage larger parcels more easily than their forebears.

The era of land consolidation in the vicinity of Moccasin Lake Ranch began with T. D. Johnson. Johnson, his wife, Mary, and their daughter Verna had lived in Montana, where T. D. was known for his generous nature. Apparently, Johnson one day loaned money to a man who claimed to be en route to British Columbia to try his luck prospecting. The prospector promised to split any of his mining profits with Johnson. Sometime later, Johnson was notified that a Canadian bank account had been opened in his name and funded with a deposit made by the prospector.

Johnson traveled to Canada, withdrew the money, and proceeded south to the Methow. It was these funds that T. D. Johnson most likely used to buy out George Thompson and expand Thompson's original holdings. After purchasing Thompson's 160-acre tract in 1902, Johnson later that year bought Pleas Rader's homestead, doubling his holdings. At the end of the year, Johnson's land extended from the Twin Lakes Road Y to the foothills just below Moccasin Lake.

During the next two years, the Johnsons acquired additional properties on the east flank of Patterson Mountain. Some acquisitions were made in the name of Verna; others in the name of Verna's husband, Earnest Williams; and still others in the name of either T. D. Johnson or Mary Johnson. In 1905, the Johnsons made an acquisition of considerable importance to subsequent owners of the ranch. They claimed the parcel of land that included Moccasin Lake and its surrounding hills. From then on, the consolidated properties were known as Moccasin Lake Ranch.

Over the next twenty years, the Johnsons opportunistically bought other parcels from their neighbors or from governmental agencies. By the early 1920s, the family had assembled eight hundred acres south of Twin Lakes.

As the Johnsons amassed their holdings, the Methow drought continued to plague dryland farmers. Many late arrivals confidently, but unwisely, claimed land on uplands where surface water was hard to find. These settlers had to rely on hand-dug wells, snowmelt, and rainfall. As the drought continued and crop production shriveled, many homesteaders tried to sell out or simply abandoned their claims and their cabins. (One such homestead cabin, built around 1909 on the prominence known as Wandling Mountain, is still standing at Moccasin Lake Ranch, in a state of disrepair. It is our intention to renovate and preserve it to give future generations a glimpse of how early Methow dryland farmers lived.)

Although T. D. Johnson was the first to aggregate land in the vicinity of Moccasin Lake Ranch, Frank Morse took land consolidation a big step further. Morse was born in Winnebago, Minnesota, in 1896. Frank and his wife, Bertha, were living in the Methow Valley with their two oldest children, Dorothy and Bob. Frank had worked in the orchards around the Gold Creek area during the early 1920s, but he and Bertha were industrious and eager to build up a ranch of their own.

By the time the Morse family first came to Moccasin Lake Ranch, T. D. Johnson had spent more than twenty years developing the property. Johnson, apparently ready to retire to a less demanding way of life, sold his land to Morse in 1925. During the next thirty-five years, the Morse family expanded Johnson's holdings from a sizable family farm into one of the Methow Valley's preeminent agricultural enterprises.

The economic environment in which Morse sought to build a ranch was definitely not encouraging, especially for a young man with a growing family, limited experience, and little access to capital. Morse's aspirations stood in stark contrast to the specter of bank foreclosure and retrenchment that hung over the valley. But longtime valley resident Carl Miller remembers the Morse family as hardworking, always striving to improve the land and with a knack for increasing its productivity.

The drought was entering its eighth year as Morse took over Moccasin Lake Ranch. He was well aware of the importance of having water rights and control of water sources. Thanks to George Thompson and T. D. Johnson, the ranch had both. Thompson's claim brought with it the right to water from Thompson Creek. The ranch under T. D. Johnson also gained rights to water from the Wolf Creek Reclamation District (WCRD). This district diverted water from Wolf Creek, routed it through a series of ditches, beaver ponds, and creeks, and stored it in Patterson Lake. From the lake, the district's water was then routed around the north side of Patterson Mountain, eventually arriving at the Twin Lakes Road Y along the northern border of Moccasin Lake Ranch.

Morse's right to water from both Thompson Creek and the reclamation district was challenged in Okanogan County Superior Court in 1926 by the state of Washington. The state argued that because of the subterranean aquifer between Patterson Lake and Thompson Creek, Morse should not be able to withdraw water from both sources, since in reality they were one and the same. Morse and others contended that their rights to reclamation district water should not be affected by any other water rights they might hold. The court ultimately agreed. In a 1928 decision, Judge William A. Huneke of the appellate court amended an earlier finding. Huneke's finding allowed Morse to eventually build an elaborate irrigation system using all the water within his rights from both sources.

Frank Morse moved quickly to expand his holdings. He made five acquisitions during his first years on the ranch. The first came in 1926, when he bought the land that had been purchased by T. D. Johnson's daughter Verna from Pleas Rader. Morse's second acquisition came in 1929, when he bought the Eric Erickson ranch. This ranch contained a substantial spring and several beaver ponds east of Patterson Lake. Erickson's property propitiously included a wooded ravine, known as Elbow Canyon, through which upper Thompson Creek flowed.

The Erickson acquisition proved to be strategically important for two reasons. First, it provided the ranch with additional access to Thompson Creek. More important, an underground aquifer flowed through it, carrying water from Patterson Lake to Thompson Creek. Without this subterranean aquifer, Thompson Creek would be dependent on winter snowfall and spring rains, sometimes unreliable sources. Happily, Morse discovered a few years later that he could pipe water to Moccasin Lake directly from Elbow Canyon over the land acquired from Verna Williams. Thus Morse's first two acquisitions were crucial in assuring a consistent supply of water to Moccasin Lake, which became the main irrigation reservoir for the ranch.

Morse took advantage of yet another opportunity to expand his holdings in 1929. The E. L. Boyd property on his eastern boundary had become available. Boyd had farmed this land for some twenty years and was now ready to retire. The purchase added 220 acres to Morse's holdings, including irrigated bottomlands near the rodeo grounds and the dry upper benches of Wandling Mountain.

Morse was not through expanding. In 1932 the heirs of Eliza Nordyke sold him her 120-acre parcel on the southeast end of Moccasin Mountain, next to the former Verna Williams land. The new piece overlooked Moccasin Lake, contained at least one seasonal water hole, and included good high-elevation rangeland. And in 1939, Morse bought a forty-acre parcel from Robert E. Williams (no relation to the H. H. Williams family) at the base of Patterson Mountain. This parcel, west of Patterson Lake Road, was later turned into an irrigated apple orchard.

By the end of the 1930s, the Moccasin Lake Ranch land consolidation under Morse was complete. He had more than doubled his holdings from the original 800 acres purchased from T. D. Johnson to over 1,800. Some 200 acres were under irrigation. In addition he leased from the government an adjacent tract of 1,680 acres on top of Patterson Mountain.

The Morse string of acquisitions was part of a consistent family plan of growth. According to Morse daughter-in-law Teddy Morse, the entire family, including in-laws, participated in all facets of the business. "Everyone was included in making farm decisions," she said, and each family member contributed to the effort in accord with his or her abilities. As Teddy Morse recalled, the strategy was to use every available dollar to purchase modern equipment or increase land ownership. As far as possible, acquired land was then irrigated and fenced for maximum productivity. Borrowing to finance land acquisitions was acceptable only as a last resort.

This approach worked well for the family through the drought years and the depressed economy of the 1930s. During their remaining twenty years on Moccasin Lake Ranch, Frank and Bertha Morse turned their attention to crop diversification, extending the irrigation system, expanding their sheep herd, and otherwise improving the ranch that had become a significant presence in the Methow Valley.

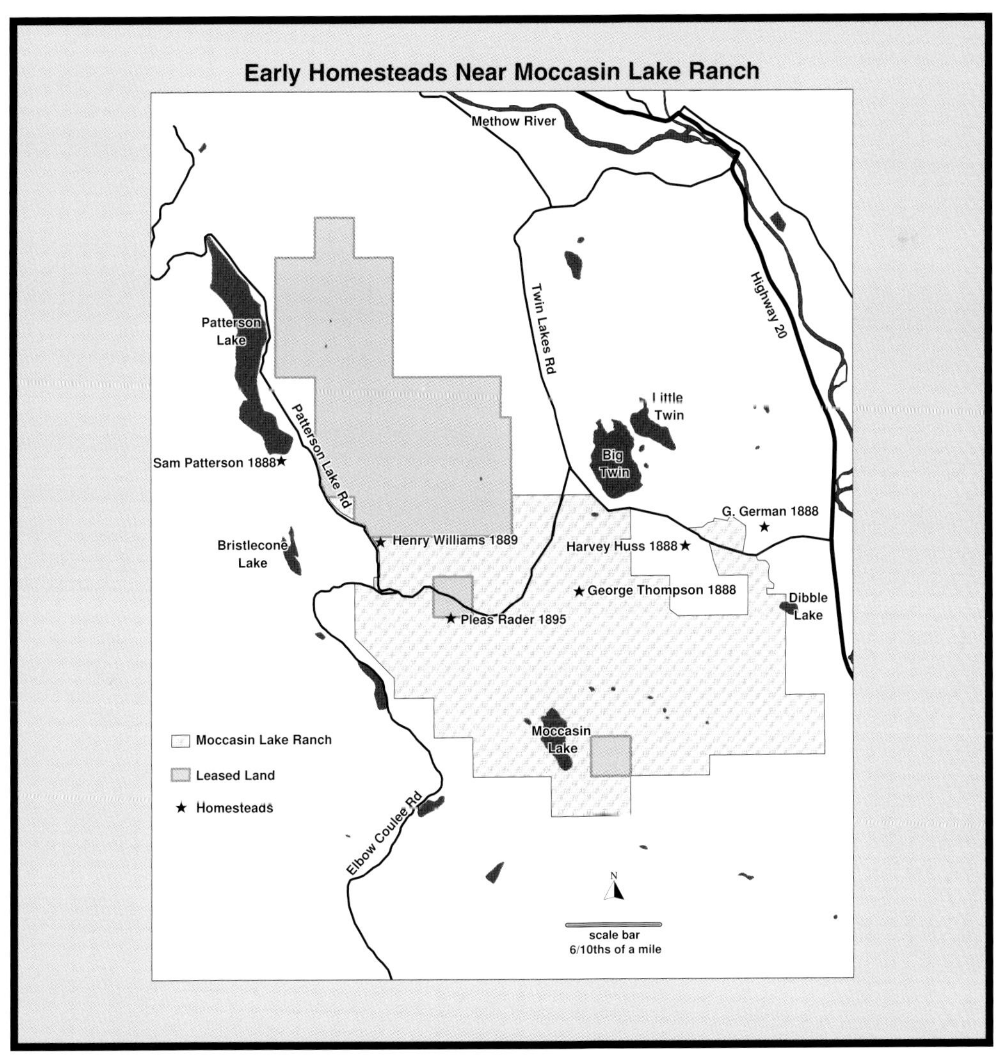

Moccasin Lake Ranch provides native vegetation for wildlife and both dry and irrigated rangeland for horses and cattle.
Photo by Molly Beck

Chapter 10

Thirty-Six Years of Ranching

Water and sheep were economic drivers for the Frank Morse family as they built Moccasin Lake Ranch into one of the Methow's finest.

Frank Morse and his son, Bob, produced an engineering marvel when they built the irrigation system at Moccasin Lake Ranch. During their early years on the ranch, father and son constructed miles of ditch across the length and breadth of the property, using the most basic of tools: picks, shovels, and horse-drawn scrapers.

At that time, electric pumps were still not available in the rural Methow. These were the days of gravity-fed flood irrigation. This meant that a network of irrigation ditches had to be constructed above the producing fields so water could be spilled out of the ditches into troughs (or rills) that carried the water across the fields to be irrigated. Careful routing of the ditches was essential to ensure a gradient that was neither too steep nor too flat. The gradient determined the water's flow rate, which in turn determined the amount of water that ended up in each field: too steep, and insufficient water was siphoned out of the ditch; too flat, and not enough water flowed to the next field.

What made the Morses' irrigation system all the more impressive was that they had neither a transit nor other surveying instruments. Bob's wife, Teddy, and her sister-in-law, Frankie Waller, remember how Bob could unerringly eyeball a spot on an opposing hillside to locate exactly where the ditch should intersect it. Although no longer in use, the Morses' work remains visible today in the form of well-shaped ditches that cross the ranch's foothills and fields.

While Frank and Bob were developing the irrigation system and clearing or fencing newly acquired land, Bertha Morse began building a sheep herd to graze on acreage that was too steep to till. She started with a few orphaned lambs called bummers. They were given to her by Jack McCall, a local sheepherder who was too busy to care for them. She bottle-fed the lambs and tended them diligently, while the herd multiplied.

In a few years there were too many sheep to keep on the home ranch, so Bertha looked elsewhere for summer and winter keep. Starting in 1930, the Morses drove their sheep every summer to public lands in the Harts Pass area of the North Cascades. In the fall, the growing herd was trailed down the valley to winter range in the Columbia Basin. In 1936, when the herd had grown to some seven thousand ewes, the Morses bought a ranch east of Wenatchee to winter the animals. They called it Sheep Camp. By 1940 the herd, which was raised primarily for its wool, numbered almost ten thousand.

During the late 1930s, Frank Morse made a critical business decision. He had a firm conviction that the market for wool was unreasonably depressed. Accordingly, he warehoused the fleeces rather than selling them at low prices. He sold the male lambs (the bucks) for meat, but kept the ewes to rebuild the herd. Then in 1942, the advent of World War II brought a financial windfall for the Morse family. The government needed wool to make millions of military uniforms, blankets, and other items. The fleeces that Frank Morse had been stockpiling were now in demand. The moment had arrived to cash in on all the hard work and planning of the preceding years. Morse sold his entire warehouse of fleeces to the government and, for the first time, had cash to pay off some borrowings and start major improvement projects. After the war, Morse sold the entire herd and got out of the sheep business altogether.

Morse soon built several needed structures. In 1944 he put up a root cellar and large hay shed on a prominence in the middle of the ranch's main hay fields. With the earlier advent of mechanized tractors, balers, and trucks, the old practice of field-stacking loose bundles of hay into enormous haystacks had been discontinued. Instead, newly cut alfalfa, after drying on the ground, was baled in the field, loaded

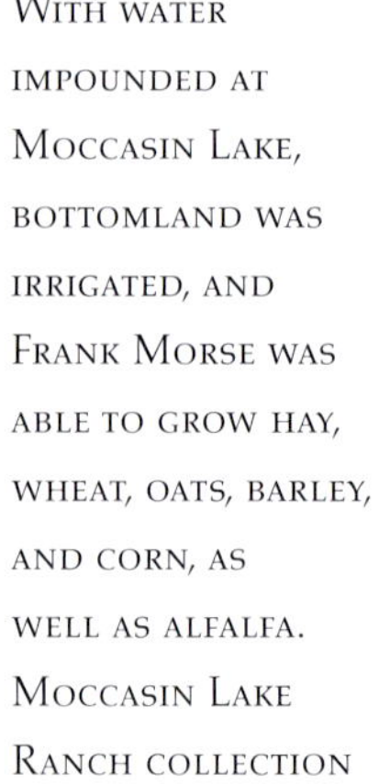
With water impounded at Moccasin Lake, bottomland was irrigated, and Frank Morse was able to grow hay, wheat, oats, barley, and corn, as well as alfalfa. Moccasin Lake Ranch collection

onto wagons, and taken to covered hay sheds. Hay stored under cover and out of the weather brought more money than hay that had been rained on, a point that did not escape the Morses.

The following year, Morse built a granary alongside his new hay shed. Wheat, oats, or barley was stored in the granary until it was fed to the livestock or sent to the flour mill. In 1947 a corn silo was added on the same promontory. By then Morse had diversified the ranch far beyond alfalfa. His crops now included corn, grain, and potatoes.

Morse made a deal with the Nalley Company of Tacoma, Washington, to provide it with high-grade potatoes to produce potato chips. These were a high-profit product. He built a sunken potato cellar with concrete floor and earthen sides. The potatoes were harvested precisely at the time when their sugar content would ensure the highest-quality chips. They were handled with great care because bruised potatoes were unacceptable.

Morse also planted an apple orchard about this time on the Williams property that he had bought at the base of Patterson Mountain. The decision to put in the orchard may have been influenced by the success of the nearby Duffy and Sandstrom orchards on Twin Lakes Road. These orchards had produced a wide variety of fruits, including apples, cherries, and plums, even apricots and peaches. (It was only after the great winter freeze of 1968 – 69 that many orchardists reluctantly concluded that the upper Methow climate was no longer suitable for growing fruit and subsequently removed their orchards.)

By far the most ambitious project undertaken by the Morses was building a pipeline to carry water from Thompson Creek in Elbow Canyon to Moccasin Lake. The lake served as their reservoir for irrigation water. What made the project so challenging was that the average fall, or gradient, over the entire two-mile route was less than one-half of 1 percent. This left no room for error if the water was to flow by gravity from creek to lake.

Bob and Frank chose to construct a pipeline rather than use an open ditch for conveying the water. A pipe would have less friction than an open ditch, and would eliminate water losses through seepage. From Elbow Canyon their twelve-inch-diameter concrete pipe wound along steep sidehills and through dense groves of conifers and then traversed open fields to the lake's western shore. Teddy Morse remembers the lusty cheers that rang out on that day in 1945 when water first spilled from the end of the pipe into Moccasin Lake.

Five years later the Morses decided to increase the capacity of Moccasin Lake by building a dam along its west side. The lake would then be able to hold more water for irrigating the expanded acreage now under cultivation. The Morses completed the dam project in 1950, raising the lake's surface some five or six feet. They also built a road around the lake that year to improve access to pastures to the south and east.

To round out his modernization program, Morse built a storage and repair shop in 1950 for his increasingly mechanized farm equipment. All these projects transformed Moccasin Lake Ranch from the labor-intensive farm of the Johnson era to a more modern commercial enterprise. By the early 1950s, the Morse family had created a self-sufficient, diverse agricultural operation equipped with modern farm machinery. The ranch had an extensive irrigation system and enough water to survive drought years. It was not dependent on any single crop. During their thirty-six years of operation, the Morse family had established Moccasin Lake Ranch as one of the principal ranches in the valley.

When Frank and Bertha Morse arrived at Moccasin Lake Ranch in 1925, their first home was a simple cabin built early in the century by T. D. Johnson. The cabin was on Thompson Creek, less than a quarter-mile upstream from where George Thompson is believed to have built his "one-room shack" in 1887. After Bob Morse was married, Frank and Bertha built a new house for themselves in 1941 a short distance away from their original cabin. Frank and Bob Morse then rebuilt the older house for Bob and his wife, Teddy.

Frank and Bertha's new house soon became headquarters for Moccasin Lake Ranch, as well as a landmark in the area. The two-story structure included three upstairs bedrooms and a large downstairs walk-in freezer where beef, venison, pork, mutton, and other perishables were stored. The downstairs also included a kitchen and dining area where meals were prepared and served to field crews during the harvest season. This house, with utilities updated and some remodeling, is still in use at the ranch today.

The Morses used another cabin, located upstream from their original home, as a residence for field workers and their families. It had been built around the turn of the century by either Pleas Rader or T. D. Johnson. Longtime Morse field hand Jimmy Stevens lived there for many years with his mother, who was known to the younger Morses as Aunt Molly. The J. Bly family then lived in the cabin for several years, followed by Alva and Belva Mundy and their four boys. All of these families helped in the fields in lieu of paying rent.

The second Frank Morse family home was built in 1941. Moccasin Lake Ranch collection

Bob and Frank Morse renovated the Mundy cabin in 1946 for Marie Morse and her husband, Mac McCall, when the couple returned to the ranch after World War II. Mac was the son of Jack McCall, who years before had started the young Bertha Morse in the sheep business by giving her a few orphaned lambs. Now Jack McCall's son had married into the family, and the young couple moved into their newly remodeled quarters. (The cabin is the oldest building still in use today on the ranch. We often call it House No. 3.)

Marie and Mac left the Methow in 1952. Bob and Teddy Morse departed two years later. By 1961, when their youngest daughter, Frankie, was grown, Frank and Bertha Morse had decided to sell Moccasin Lake Ranch and retire to Sheep Camp, near Wenatchee. The couple had a long period of retirement before Frank died in 1992, and Bertha in 1997. The three oldest Morse offspring also died during this period, leaving only Donna and Frankie along with their sister-in-law Teddy to recount their experiences at Moccasin Lake Ranch. During their years on the ranch, the Morse family witnessed both the good times and the bad. Their many accomplishments are still visible, and have contributed to the success of today's operations.

The Winthrop bridge, with Winthrop School in the background. Dick Webb collection

Chapter 11

Hints of Things to Come

Better roads, the ready availability of cars, and abundant fuel to power them brought outdoor enthusiasts to the Methow, where they discovered a rural wonderland.

Through the 1930s, the economy and social fabric of the valley still centered on the family farm. Yet two decades later, when Frank Morse put Moccasin Lake Ranch up for sale, agriculture in the Methow Valley was in a state of steady decline. During the intervening period, a series of events changed the Methow's culture and lifestyle forever.

The first and most dramatic development was the U.S. entry into World War II. Virtually every physically fit young man was called to active duty, leaving the Methow Valley and its farms far behind. Unlike World War I servicemen, veterans of the Second World War were offered the chance to get a college education at government expense. Armed with college degrees and newfound interests, many third-generation Methow residents eagerly pursued higher-paying careers outside the valley, never to return to farm life. However, a few descendants of pioneering families did return to the Methow to seek careers in agriculture or mining. This group included sons of the LaMotte, Lehmann, Miller, Filer, Prewitt, Thurlow, and Stokes families.

During the first years after World War II, agriculture remained strong in the Methow. Family farms had gradually been modernized, land prices were still based on agricultural value, and the postwar economy was prospering. The advent of major tourism, still more than two decades away, had not yet inflated the local economy or the value of real estate.

Postwar farms had the benefit of low labor costs and newly available mechanized farm equipment that rapidly improved productivity. Electricity supplied by the Rural Electrification Administration in the late 1930s provided power to automate many farm chores. Cows could be milked using electric-powered suction devices. Water could be pumped from wells, and electric lights allowed farmers to extend their workdays.

The valley's north-south highway, championed by W. A. Bolinger, was finally paved and completed in 1938. Its eleven bridges across the Methow River permitted round trips from Mazama, in the upper valley, to Pateros, at the Columbia River, in less than a day. The state highway department carved a road out of the steep terrain of the Okanogan highlands, in the vicinity of Loup Loup Pass. This road, leaving the Methow Valley about three miles south of Twisp, was first made usable in 1934 and improved the following year, when it was graveled and oiled. At that time the road, state highway 20, was extended eastward to the town of Okanogan, the county seat.

The new roads greatly facilitated travel for Methow residents. They also brought an economic threat to its merchants. It was now easier for businesses outside the valley to compete with local merchants. Creameries in Twisp and Winthrop soon went out of business. The prices of locally grown fruit, vegetables, and dairy products were no longer competitive with those of produce imported from the Columbia Basin, where the growing season was longer and winters were milder. These items could now be preserved with the use of electrically powered refrigeration equipment. Many Methow residents simply resorted to consuming much of the food they raised. The ubiquitous icehouses previously found throughout the Methow were fast becoming obsolete. The past way of life was rapidly slipping away.

Most veterans of World War II who did return to the Methow found limited job opportunities. Mechanization meant that farms needed fewer workers. Local mines, still facing the twin problems of low mineral content and high transportation costs, were operating only part-time if at all. One bright spot for employment was the Wagner sawmill in Twisp. Built by Otto Wagner in 1941, it became the Methow's biggest postwar industrial employer, with as many as four hundred workers during the 1940s and 1950s. Logging crews that supplied the raw material for the mill pushed into the valley's backcountry, building new roads as they went. The loggers, in turn, were supported by a fleet of trucks and drivers.

These jobs not only generated good wages for employees, but also provided valley residents with a feeling of security. The shrill sound of the mill's whistle that signaled the start of work each morning was welcomed by residents because it represented a paycheck and a way of life. As long as the supply of pine and fir trees lasted, jobs seemed secure.

Then one day the sawmill stopped making the components, called shook, to produce apple boxes. Orchardists had begun to replace the traditional wooden boxes with lighter, cheaper plastic boxes for shipping their apples and other fruit. This technological advance was a harbinger of difficult times ahead. The Wagner mill tried to replace the lost business by diversifying into charcoal briquettes made from sawdust, but this effort met with only limited success.

Other quiet hints of the Methow future began to appear during those early postwar years. Urbanites from the Puget Sound region, with cars and gas now available and affordable, began to visit the Methow and the relatively unexplored regions of the

SEVERAL LOCAL SAWMILLS SERVED METHOW VALLEY RESIDENTS DURING THE EARLY 1900S. SHAFER MUSEUM COLLECTION

North Cascades. What they found was a pristine valley that was little changed from the one that had entranced pioneers two generations before. It was still lightly populated, with almost no industry or pollution. Visitors delighted in the light powder snow for skiing in the winters and the open range and wooded hills for horseback riding in the warm summers. They discovered fish in the valley's lakes and rivers and game on its lands.

This growing enthusiasm for the Methow as a recreational getaway received a temporary setback in the spring of 1948, when an epic flood virtually destroyed every bridge between Pateros and Winthrop. It also washed away a number of riverside homes. The spring snowmelt raised the Methow River to its highest level on record. The flood is believed to have been equal to or greater than the flood of 1894 that wiped out the little community of Silver. Merchants went for weeks without supplies, power, or phone. The only passable exit to the south during the summer of 1948 was along one of the original pioneer wagon trails over the Okanogan highlands. The valley's fragile economy was shaken, and the mettle of its residents was tested. But once again these industrious families, many of them third-generation descendants of the pioneer settlers, worked together to rebuild their lives and survive.

After the damage was repaired and new bridges were constructed, urban visitors gradually returned. This didn't escape the notice of local entrepreneurs. Some ranchers converted their spreads to dude ranches, letting city folk ride with the local cowboys and pitch in with the chores while paying for the experience. One of the first full-scale dude ranches was the Sunny M operation, originally homesteaded by Clint Shulenburger in 1889. By 1950, this property was owned by Joe Barron, who loved to entertain his guests. He encouraged his ranch hands to stage impromptu rodeos, cookouts, and evening campfires. Elsewhere, dude wranglers could also try out their cowboy skills at Dr. Blend's establishment or at Paul Spaeth's ranch, where Jack Rader, an accomplished horseman, entertained and ran the stables.

Another form of dude ranching was offered by Jack Wilson, who gained a reputation as a packer into the high country. Wilson took riders on horseback into the spectacular mountains and lakes of the North Cascades. There they slept out under the stars while the campfire burned low. Wilson developed such a devoted following that he constructed several cabins at his ranch near Mazama for year-round use by campers and outdoorsmen. These cabins, now renovated, are still in use as part of the Freestone Inn.

During this time the valley's winter activities were promoted by skiers, who found snow that was drier and more consistent than that on the west side of the mountains. Skiers were further enticed by the solitude and the untrammeled expanses of snow-covered state and federal lands for cross-country (Nordic) skiing. Some of the veterans who returned from the battlefields of World War II were men who had joined the Army's famous Tenth Mountain Division, a group of expert skiers who trained to fight in the Italian Alps. Included in this group were Methow Valley residents Warren West, Howard Brewer, and Buck McKinney. They shared a love of the outdoors and mountaineering. Surely they helped spread the word about the wondrous winter snow conditions in the Upper Methow. Gradually a few bed-and-breakfast operations emerged to join some renovated inns in offering lodging during both the winter and summer seasons. By the mid-1950s, a fledgling tourist trade was taking hold.

Still to come was the development that would alter the Methow Valley lifestyle perhaps more than any other. Over the years the state highway commission had been intrigued with the concept of a northern cross-state highway—an east-west route that would traverse the majestic North Cascades. The idea dated back to the early nineteenth century, when Alexander Ross attempted to find an overland route to the sea. Then in 1883, First Lieutenant George Bacus was ordered by General Sherman to survey the North Cascades to determine if there was a usable pass to the Pacific Coast. Still later, mining interests considered building a road from the Slate Creek mines to tidewater smelters. There were even surveys made for a proposed railroad, but when mining declined, interest in this venture waned along

Six of Jack Wilson's original cabins at Early Winters Resort at Mazama were refurbished in the late 1990s and became the site of the Freestone Inn. Photo by Al W. Dorow

with it. After several false starts, the cross-state highway proposal took on momentum as politicians began to sense a growing public interest in having convenient access to undeveloped wilderness. The North Cascades was truly such a place.

In contrast to later periods, when it became popular to leave wilderness areas roadless, the prevailing thinking at the time was that a highway into the North Cascades would be beneficial. It would open the wilderness for public recreation, provide construction jobs for residents along the route, and stimulate the Methow's languishing economy. In 1956 the state's highway commissioners rode through part of the proposed route on horseback. The outing proved to be decisive. The commissioners were impressed by the spectacular scenery. Surely a new road would stimulate the economy of both the Methow and the entire county. Soon the state highway department started detailed planning for a North Cascades highway, a route that would become the instrument of immense change in the Methow Valley.

Covered in snow, Moccasin Lake Ranch looks much as it did in 1960, when Jon Titcomb first saw it. Moccasin Lake Ranch collection

Chapter 12

A New Vision

In the process of striving to make Moccasin Lake Ranch financially self-sufficient, Jon Titcomb invested in modern farm equipment and techniques, new buildings, and a herd of Charolais cattle, an exotic European breed.

When Jon Titcomb first visited the Methow Valley in late November 1960, it was under almost two feet of snow. He was there with real estate agent Bert Wassell to check out a property in the upper valley. It was the latest effort in Titcomb's search for a ranch to serve as a project for his retirement years and a rural retreat for his family. Once there, he concluded that the property didn't suit his purpose. As they began driving south out of the valley, Wassell remembered a ranch near Twin Lakes that might come on the market. They decided to visit the place and turned off the main road just south of Winthrop. They soon were looking over the lower fields of Moccasin Lake Ranch, its positive attributes recognizable even under the blanket of snow.

Titcomb returned to the Methow in early January and met with Moccasin Lake Ranch owner Frank Morse. The men were five years apart in age and had vastly different backgrounds, but they shared an interest in agriculture and the outdoors. Titcomb, age fifty-nine, was not a novice at farming. Jon had spent his early married years working at Weyerhaeuser but raising his family on a small farm near Everett, Washington. While ascending the corporate ladder, he had managed to find time for routine farm chores both before and after work at the office. Later in his career, after the farm had been sold, he often spent weekends and vacation time in the wilderness hunting or fishing. His closest partners were generally the Weyerhaeuser foresters, those who knew the deepest secrets of the backwoods on company lands.

Morse, age sixty-four, was ready to retire from active ranching after more than three decades building his farm and sheep ranch. He had started as a field worker in the Methow's apple orchards and risen to ownership of Moccasin Lake Ranch. He had survived the drought years, borrowed and repaid money during the Great Depression, overcome a challenge to his water rights by the state of Washington, and built with his wife, Bertha, a sheep herd to ten thousand head. Now, the sheep had been sold and his children had grown and moved away. It was time to turn the ranch over to other hands.

Morse happily described the ranch's features to his potential buyer. He told Titcomb that the ranch was big enough to be economically self-sufficient. The sandy soil permitted growing a variety of crops with excellent yields. The lake's elevation was sufficient to irrigate croplands without the need for electric pumps, an important cost-saving consideration. Finally, the proximity of the Methow Valley to the Puget Sound basin would make it easy to get to the ranch from Seattle and other cities of Western Washington.

These attributes struck responsive chords in Titcomb, who also recognized several additional virtues. The ranch's natural beauty evoked fond memories of his youth in the wilderness of Vermont. The property was sufficiently large to serve as a family retreat. And, of considerable interest to this lifelong fisherman, Moccasin Lake might well be suitable for top-notch trout fishing.

Titcomb sat down and wrote out a purchase option before he left that day and both men signed it. They reached agreement on all remaining details over the next few weeks, and the sales transaction was recorded on March 9, 1961. Jon Titcomb became the fourth owner of the land that was originally homesteaded three-quarters of a century earlier by George Thompson. The property was expanded and improved by T. D. Johnson; then Frank Morse added more land and developed an irrigation system; now Jon Titcomb would build on this ranch tradition of robust growth.

After buying Moccasin Lake Ranch, Jon Titcomb was in his element, eager to put his philosophies of farming into practice. One of these philosophies involved hands-on management. Accordingly, he took an interest in every aspect of the operation. No detail was too small for Jon to review. During the first winter he helped design time sheets for the hired crew to keep track of their work hours. He discussed the best location for the wood-burning stove in the new shop he planned to build. He reviewed fertilizing procedures for potatoes with purchasing executives from the Nalley Company. He even had time to plan for future irrigation ditches, noting that their gradient should be no greater than three inches for every one hundred lineal feet.

Another of Jon's philosophies was that nothing should go to waste. He felt that table leftovers should be fed to the pigs rather than thrown into the burn barrel. Accordingly, a couple of sows soon took up residence in a pigpen that Jon had built at the ranch. Scraps of plywood were used to build the shelter and, of course, at the right time the animals would be converted into hams and bacon.

Jon was pleased to contemplate incorporating yet another of his philosophies at the ranch — one that looked favorably at opportunities to combine outdoor recreation with business. As a boy, Jon was exposed to this concept while traveling with his father, whose responsibilities as game commissioner in New York State sometimes necessitated wilderness expeditions that would involve camping and hunting. During his business career at Weyerhaeuser, Jon was occasionally able to include some fishing on streams that flowed through company forests. At Moccasin Lake Ranch, opportunities would certainly arise to fish and hunt while overseeing daily operations. Jon looked forward to such occasions.

As the winter snows melted away that first year and uncovered the fields, it became apparent that there was a major discrepancy between the acreage that Frank Morse had claimed was available to irrigate crops and what was actually available. This was troubling to Jon not only because the purchase price was based in large part on the amount of irrigated land, but also because, in the semi-arid Methow climate, both alfalfa and potatoes required irrigation to be economically viable. Jon argued the issue with Frank Morse and real estate agent Bert Wassell, but without satisfaction.

Titcomb wanted a major price adjustment to offset the extra investment that would be needed to bring the irrigated acreage to the planned level. Talks between Titcomb, the thrifty Vermonter, and Morse, the shrewd farmer, became increasingly heated. By the end of 1962, Titcomb had decided to take Morse to court. Lawyers were engaged, but it would be the spring of 1964 before the case would be heard by a judge.

In the spring of 1961, Titcomb set about hiring his ranch team. First, he hired Harley (Hod) McKay as the ranch's foreman. Hod and his wife, Mary, moved from Kent, Oregon, into House No. 1 (Frank Morse's main house). In April a second employee, Gus Kopke, was hired and moved with his wife into the house on Thompson Creek previously inhabited by Bob and Teddy Morse. Before his first year as owner was over, Jon hired a third hand for seasonal work, a fellow known as Tex.

Tex was one of the more colorful ranch hands to ever work at Moccasin Lake. Family members still remember the night of a great electrical storm in 1962. A torrential rain was repeatedly punctuated with brilliant flashes of lightning. They struck ever closer to the old Erickson house, where we were all staying, at the top of the grade to Patterson Lake. Each flash lit the house as if all its lights had been switched on, turning the darkness of night into midday brilliance. At the height of the storm, Tex arrived with a great flourish. He topped off his arrival by waving a six-shooter in each hand and exclaiming that there was nothing to fear because he was on hand.

We survived the storm without further incident, but the lightning set off a brush fire in the vicinity of Elbow Coulee. Nim Titcomb, Jon's only son and a licensed civil engineer, was surveying the ranch's southern boundary when he noticed smoke

Roy and Marge Simons. Roy replaced Hod McKay as top hand at Moccasin Lake Ranch in the mid-1960s. Moccasin Lake Ranch collection

coming from the far side of Elbow Coulee. By that time the fire had already engulfed several acres of timberland. The ranch's Caterpillar D6 dozer was quickly brought in to clear away brush in the path of the flames. Professional firefighters soon joined the battle. Family members and neighbors grabbed shovels and picks to heap dirt on smoldering embers. It was an exciting but sobering experience that increased our awareness of the many hazards of rural living.

Jon Titcomb plunged into the affairs of the ranch. He was clearly less a retiree than a man who had simply changed careers, from corporate executive to modern-day rancher. The projects that he and his ranch hands undertook in his first three years as owner were not those of a man looking to take life easy.

First came the machine shop. Jon understood that as automation replaced the labor-intensive farming of earlier times, the ranch would need a place to maintain and repair its tractors, haying equipment, cultivators, and other contrivances. A large shop with overhead hoist and concrete floor was erected during the summer and fall of 1961. Large sliding doors admitted the widest farm equipment. Windows on the northwest wall let daylight flood the shop. Mechanized farm equipment was parked in a covered shed built next to the shop. Jon was not extravagant in equipping the ranch, but his corporate experience had taught him that automation was a fact of life in business — and he could see that farming was increasingly a business as much as it was a way of life.

Jon made an important advancement in ranch irrigation by installing pressurized sprinkler systems to replace Morse's flood-irrigation method. Water from Moccasin Lake was now routed to a newly constructed concrete head box located high on an adjacent sidehill. Screens in this structure removed waterborne debris and vegetation before the water spilled into buried main lines that led to the lower fields. On the sides of these fields were smaller buried pipes, with risers to which portable hand lines with sprinklers were connected. The new system improved water efficiency and increased alfalfa production. All available croplands on the ranch were soon sprinkler-irrigated, using water from either Moccasin Lake or the Wolf Creek Reclamation District. Over time, self-propelled wheel lines gradually replaced the more labor-intensive hand-moved sprinklers on the larger fields.

After potato prices had fallen to a new low, Jon got out of the potato business to concentrate on raising cattle. The ranch's main customer, the Nalley Company, could buy potatoes more cheaply from producers in the Columbia Basin. Titcomb's strategy now was to establish a cow-calf herd of 600 to 700 head. He planned to grow enough alfalfa on the ranch to feed this herd during the winter when snow precluded grazing. He started in the spring of 1962 by purchasing about 250 head—some steers for fattening and resale, but mostly bred heifers that would calve the following spring and thus build the herd's numbers. By 1964 the herd had grown to 350 cow-calf pairs.

Jon realized it would be necessary to enhance the ranch's range to accommodate the expanding cattle herd. This involved finding surface water, building fences to contain the cattle, and improving the range pastures. The drought of the 1920s and early 1930s was long since over, but many upland pastures lacked their own ponds or other surface water. In one innovative scheme, Jon considered installing a windmill near the top of Patterson Mountain and using it to power a well pump. He corresponded with windmill manufacturers, but in the end, he dismissed the scheme as impractical.

He built three large hay sheds to replace the older ones from the Morse years. One of these sheds housed a feed mill to mix winter feed rations for the cattle. Conveyors moved molasses, grain, chopped silage, and vitamins to an elevated mixing bin. The mixed rations were released down a chute into a feed wagon parked below. Each winter morning, the wagon would be towed by farm tractor to the fields where the cattle were waiting.

By 1961, housing at the ranch was most certainly outdated, with the sole exception of Morse's main ranch house. Jon intended for his ranch manager and family to live in that house, while other ranch hands would stay in the older cabins. Jon planned to spend considerable time at the ranch along with his wife, Gaye, so he made some modest improvements to the house that had been built in the mid-1920s by Eric Erickson. Inconveniently, that house was located almost three-quarters of a mile from the main ranch, at the top of the grade to Patterson Lake. When Gus Kopke left the ranch and newly hired Roy Simons took up residence in Kopke's old house, it provided an opportunity for Jon and Gaye to move into the larger house where Roy and his wife, Marge, had been living. This house, called No. 3, was better situated to oversee ranch affairs because of its proximity to the shop and other ranch buildings. The kitchen's wood-burning stove was now replaced by an electric stove with oven and trash burner. The primitive icebox was converted to an electric refrigerator. The living room was expanded and fitted with a fireplace and a large view window. Still later, a deck around the south and east sides of the house was added, affording a fine view across the lower portion of the ranch. House No. 3 became ranch headquarters for Jon and the place where the entire family stayed on visits until the mid-1990s. (It is still used by Jon's daughter Greata Beatty and her family.)

Early in his ownership, Jon began working with the State Game Department on a program to stock Moccasin Lake with rainbow and eastern brook trout. The first step was to remove the trash fish. Then several hundred small trout were planted in the lake, during the spring of 1962. The state took care of stocking the lake, and in return Titcomb agreed to let the public fish it. Moccasin Lake was opened to public fishing in the spring of 1963. Only fly fishing with barbless hooks was allowed. Many regulars were vocal in enforcing the rules. In fact, anyone who violated the rules could expect to be admonished by fellow fishermen who appreciated the

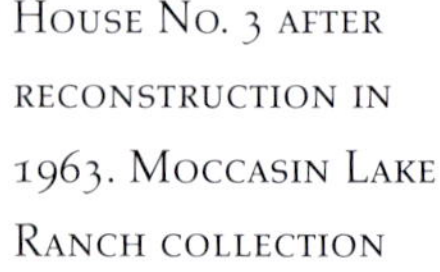

House No. 3 after reconstruction in 1963. Moccasin Lake Ranch collection

program. Although the trout were small in the beginning, enthusiasts who made the one-mile hike to the lake from the ranch's parking lot showed their feelings by returning several times. They understood that the fish would grow as time went on.

With three years of ranch operations completed, Titcomb was anxious for his lawsuit against Frank Morse to be heard in court. He accumulated a great deal of evidence to substantiate that the actual number of acres that could reasonably be irrigated was considerably lower than he had been led to believe. He was convinced that Frank Morse had overstated the irrigated acreage and thus had been overpaid for the land. Jon not only had the entire ranch surveyed, with detailed calculations of the acreage in each irrigated field, but he also hired an expert to independently verify the survey using measurements made from aerial photos.

In the spring of 1964, the case went before a judge in Wenatchee. A few months later, the judge concluded that, although Morse's tabulation of irrigated acreage was undeniably exaggerated, there was no evidence that Morse had ever guaranteed the exact amount of land that was or could be irrigated. Irrigated acreage, according to the judge, had been only a discussion point during the negotiations rather than a contractual matter. On the other hand, the judge recognized that it would be expensive to expand the existing irrigation system to the extent that it could irrigate the acreage Morse had claimed. He therefore ordered that the purchase price for the ranch be reduced — but the reduction was to be less than 5 percent of the original purchase price. The amount stipulated was far too small in Jon's view.

Jon worried that the ranch would not produce enough alfalfa to pay for itself with the existing irrigation system. He was also aware that building the cattle herd was a long-term and expensive project. After months of agonizing, he decided to sell Moccasin Lake Ranch. In October 1964 he ran his first ad in the *Wall Street Journal*. There was little response. By the end of 1965, Jon came to accept what he felt was a disappointing outcome to the Morse lawsuit. He abandoned any further attempts to sell the ranch.

Jon now began to consider getting into the purebred cattle business. In parallel with building the commercial cow-calf herd, the ranch would also build a herd of registered purebred cattle. Purebreds would each bring considerably more money than their unregistered counterparts. Purebred bulls and heifers would be sold to breeders who were expanding or upgrading their own herds. Sales prices for these animals were determined by their bloodlines and genetics rather than weight. In contrast, calves from the commercial herd would be sold to feedlots on a weight basis. Jon believed that this new strategy would increase revenue and help make the ranch economically viable.

In 1965, the large and distinctively blond Charolais was Titcomb's breed of choice. Moccasin Lake Ranch collection

He evaluated several breeds of cattle for his purebred herd before choosing the Charolais breed, which was popular in France and was gaining considerable attention in the United States. Many cattlemen believed that the traditional English breeds, Herefords (also called whiteface) and Angus (both the black and red varieties), were being surpassed in various important measures by the new so-called exotic imports: Charolais, Simmental, and Limousin. These breeds, imported from the European continent, were reputed to produce larger calves and ultimately more meat. They not only grew bigger than the English cattle, but they also grew faster.

Jon eventually selected six heifers and ten bulls from the McElhaney Cattle Company in Tempe, Arizona, to inaugurate his Charolais herd. In April 1965, the animals were loaded aboard a railroad boxcar that was partitioned to keep the heifers and bulls separated. Water troughs at each end of the car and the feed bunk in its middle were replenished at stops during the trip from Tempe to Spokane. But misfortune struck along the way when two of the animals were seriously injured in the boxcar. The accident resulted in the death of one of the bulls and caused permanent impairment to one of the heifers. It was an inauspicious start to the new program. In spite of this setback, the purebred cattle program grew and thrived over the next few years.

To increase revenues, Jon needed more calves to take to the fall auctions. He couldn't do this, of course, without carrying more brood cows through the winter. To make this work financially, he considered wintering his cattle where there was warmer weather and less snow. Such a place would shorten the feeding time and thereby reduce the cost of feed. Jon visited several farms in the Columbia Basin and found both accommodating weather and cheap feed readily available. However, he also found that the cost of transporting the cattle to and from the wintering grounds, along with the cost of pasture rental, would more than offset the savings in feed cost. Jon was forced to conclude that wintering the cattle in warmer regions did not solve the basic cash-flow problem caused by low market prices for livestock.

As the purebred herd grew in numbers, it became evident that Charolais consumed a disproportionate amount of feed. They consumed at least 15 percent more feed each day than the commercial cattle. To make matters worse, the young Charolais bulls and heifers needed extra daily supplements of grain, molasses, and vitamins to maintain the growth rates that would make them more desirable to prospective buyers. Given the high cost of feeding them, their much vaunted size advantage was beginning to look like a mixed blessing.

To compound the problem of making the ranch economically self-sufficient, Jon discovered that it was short of range for summer grazing. He recognized that both winter feed and range carrying capacity hinged on putting more acreage under irrigation. Jon decided to irrigate the uplands around Moccasin Lake. He also started irrigating some of the smaller croplands around the swamp. He calculated that each additional acre of cropland brought under irrigation would produce three and a half tons of alfalfa, enough to support almost two head of cattle during the winter.

More irrigated land, of course, meant a need for more water. The answer seemed to lie in raising Moccasin Lake's water level once again. Before proceeding with a new dam at the lake, however, Jon had to take into account a number of sometimes

Moccasin Lake after its second expansion in 1965. Photo by Barry Provorse

unfathomable considerations: How many additional cattle could be carried through the Methow winters? How many more cattle would the range sustain during the summer season? And most problematic of all, what prices would these cattle fetch on the local market? The numbers, though speculative, looked sufficiently promising to encourage Jon to go ahead with the project. In 1965, fifteen years after Frank Morse had first built a dam at the lake, Jon added another five feet to the original. The lake now covered approximately thirty-two acres and had a maximum depth of sixty feet.

Jon now reviewed his original objectives in buying the property. As a retirement project, ranch operations were fulfilling his need for a continuing business challenge. Making the ranch economically self-sufficient was proving to be elusive. However, another objective, that of providing the family a retreat from the pressures of everyday urban life, was being handsomely accomplished.

The fly-fishing program at Moccasin Lake was popular with both the family and the public. Also available for fishing was the nearby Methow River. After the high waters of the spring runoff had subsided, Jon liked to lead fishing trips down the Methow with one or two family members on board the McKenzie River drift boat that had been in the family for a number of years. Nice trout could be taken from the boat or from the riverbank.

Another recreational activity that both Jon and his wife, Gaye, looked forward to was harvesting apples and cherries from the fruit trees by House No. 3. Though small, the trees were prolific and supplied enough fruit to enable Gaye to bake pies for the family. Apples also were squeezed to provide juice for apple cider.

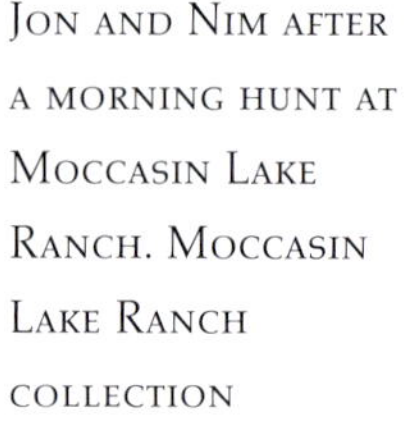

Jon and Nim after a morning hunt at Moccasin Lake Ranch. Moccasin Lake Ranch collection

Titcomb family members outward bound on the fresh, dry snow of Moccasin Lake Ranch. Moccasin Lake Ranch collection

Several jugs were filled with the juice flowing from the hand-cranked cider press that was put into operation each fall. While the bounty of the fall harvest was being processed, the grandchildren often had fun bobbing for apples in a large tin filled with water.

Jon, his son Nim, and his sons-in-law Mac Beatty and myself were also enthusiastic about hunting game on the ranch, which included whitetail and mule deer, quail, grouse, ducks, and geese. Jon enhanced the opening day of some fall hunting seasons by procuring several pheasants that were let loose in the fields for bonus hunting. On these occasions, family and friends would assemble for a few days of outdoor sport. There was always an assortment of fish and game to be cleaned and eaten and a myriad of stories to tell at the end of the day. By that time, Jon was using a cane for assistance in getting around. But he still could be seen on his horse traversing the ranch's hills and fields during both the summer months and the hunting season.

Later in the year, as winter set in, the dry powder snows provided ranch visitors with fine sledding and cross-country skiing. Sledding was especially popular during clear evenings when the stars joined the moon in lighting the snow. Although Jon and Gaye Titcomb did not participate in winter sports, they were always happy to hear about the outdoor adventures experienced by family members, whose numbers had grown to eleven by the latter part of the 1960s.

In the mid-1960s, as Jon continued wrestling with operational problems, he discovered that the ranch immediately to the east of Moccasin Lake Ranch was for sale. Here was a chance to acquire land that included tillable fields to produce additional winter feed as well as rangeland for summer grazing. The 860-acre ranch was owned by Gaylord and Edith Wandling. It had the potential of providing exactly what Jon Titcomb needed, and it was dotted with ponds and springs to water the cattle. The property also included corrals, a hay shed, a ranch house, and one half of Dibble Lake.

The Wandling property had a long and enterprising history, much like that of Moccasin Lake Ranch itself. While T. D. Johnson and Frank Morse were expanding their landholdings around Moccasin Lake Ranch, a similar aggregation of land was taking place just to the east. It started with the Garrett family and their One Bar Ranch. J. C. Garrett, a veteran of several Civil War battles as a Confederate cavalry officer, moved to the Methow Valley in 1900. That year, the senior Garrett bought the Harvey Huss homestead from David Herstine. Garrett's son Ras and a friend bought the Granderson German homestead in 1902 (site of the present-day Liberty Bell High School).

The two Garrett properties became the nucleus of the One Bar Ranch, which eventually grew to 320 prime acres extending from Big Twin Lake to the current rodeo grounds and eastward to Dibble Lake. J. C. Garrett died in 1908, and his son sold the combined properties a few years later. According to an account by longtime Methow resident and businessman Dale Dibble, the One Bar Ranch experienced a succession of owners. By 1926, when the drought in the Methow was in its tenth year, the former Garrett property had become a victim of lack of water and absentee owners. It went into default and was turned over to the Federal Land Bank of Spokane.

The bank deeded the property in 1931 to Jefferson J. Nagley and his wife, Louisa. They had filed a homestead claim in 1910 on land adjacent to the Garrett property. The Nagleys built several structures on this land while clearing and improving it. The year after receiving the deed to a part of One Bar Ranch property, the Nagleys passed along a portion of that land to their son, Guy, and two of their daughters, Phoebe and Myrtle.

Over the next twelve years the younger Nagleys developed the combined lands they had inherited. They took advantage of water from the Wolf Creek Reclamation District not only to irrigate their lower fields but also to operate a surprisingly effective ram-jet mechanism to irrigate their orchard situated on an elevated bench. In describing this system, Dale Dibble wrote, "Water from the Wolf Creek Ditch dropped down a steep hill with perhaps a fifty-foot vertical drop through an eighteen-inch pipe. This pulsating mass of water would apparently strike a cylinder, driving perhaps 10 percent of this water in a smaller pipe far up the hill onto land the Nagleys wanted to irrigate."

The Nagleys further expanded their holdings along the northeast slopes of Wandling Mountain. They eventually owned six hundred acres, a herd of milk cows, and both alfalfa and dryland grain fields.

The Nagley family sold its holdings in 1944 to Gaylord Wandling, a valley newcomer. Wandling was determined to expand them, concluding that agriculture was benefiting from the return of wet years, and water was readily available from the Wolf Creek Reclamation District. He immediately set out to buy additional lands adjacent to the former Nagley holdings.

Wandling's aggregation of land in the 1940s differs in one important respect from Frank Morse's land acquisitions at Moccasin Lake Ranch during the preceding decade. Morse's expansions almost always included surface water and accompanying water rights. Wandling, on the other hand, acquired dry upper benches. Some of these benches held ponds and springs during wet years. But most of the mountain, and Wandling's acquisitions, consisted of dry upland range.

Despite a lack of water, there were a few early-twentieth-century homesteads on these benches. The homestead cabin of Henry Crockett, who received his deed from the government in 1908, still stands today, although in badly deteriorated condition. Crockett, a bachelor, cleared rocks from nearby fields, where he then planted alfalfa and grains. Water came from a well dug alongside a pond next to the cabin. Crockett, or his successor, William Sheehy, built a threshing shed to process sheaves of grain. A derrick for stacking hay was constructed, as was a primitive wagon road from the homestead to the lower fields next to Dibble Lake. Crockett lived in the cabin until 1912. Then he sold to Sheehy — also a bachelor — who lived there and farmed the land until 1943.

The Crockett and Sheehy parcel and adjoining lands were owned by a succession of individuals. They had been willing to gamble on the availability of water but eventually found that the land could not support them. Gaylord Wandling followed up his purchase of the Nagley ranch by acquiring William Sheehy's homestead and lands that had been owned by C. H. Blanchard, Perry Brewster, Minnie Bishop, Charles Berg, Nels Peterson, and the state of Washington. By the time he put his ranch on the market in 1966, Wandling had accumulated 860 acres.

Excited after hearing about the availability of the Wandling ranch, Jon Titcomb was dismayed to learn that Wandling's real estate agent had already received an offer on the property. Jon hurriedly arranged to meet with Wandling, trying to get his support for the logical combination of the two adjacent properties. Since Wandling had not yet accepted the competing offer, he agreed to meet with Titcomb.

The Crockett place, the last original homestead on Moccasin Lake Ranch. Photo by Scott Chambers

To avoid the possibility of having to share in paying the real estate agent's commission, Jon proposed to lease Wandling's ranch until the agent's exclusive listing on it expired. The men signed a lease that allowed Titcomb to farm and graze the property until a formal sales contract transferred ownership to Moccasin Lake Ranch. That contract was signed on August 10, 1966.

With this purchase, Moccasin Lake Ranch had grown to 2,768 acres through a series of acquisitions over a period of six decades, first by T. D. Johnson, then Frank Morse, the Garrett family, the Nagleys, Gaylord Wandling, and finally Jon Titcomb. The ranch was leasing another 1,680 acres, for a total of more than 4,400 acres under its management.

Jon's health continued to decline during the latter 1960s. Pain in his ankle required him to wear a brace, and his cane had become a constant companion. He could no longer ride horses, but he was able to get around the ranch by car and truck. His interest was as strong as ever. He oversaw installation of new fencing on the Wandling property and water development on the uplands. The Wandling purchase had given him new hope that ranch economics could be made viable.

Jon could look back over his first five years and see the many fruits of his labor. A growing commercial cattle herd was grazing on the slopes of Patterson Mountain. The Charolais herd showed desirable genetics and was doing well. The irrigation system had been vastly upgraded, and solid management had been hired for the ranch. This was critical because Jon's health problems meant he needed more and more to rely on Roy Simons and Hod McKay to plan and conduct ranch operations.

The Wandling acquisition was the zenith of Jon's active involvement in ranch affairs. After that he spent more time communicating with those outside the ranch on issues important to him because they could affect life at the ranch. He wrote to Washington State's congressmen imploring them to oppose extending the minimum wage to farm employees. (Young men in rural areas had a hard enough time finding jobs, he argued, without making it even more difficult by raising wages to a level that would discourage farmers from hiring them.) He also advised congressmen to oppose extending the highway gas and diesel tax to farm equipment. (This tax was meant to be used for highway maintenance, but farm equipment was not driven on highways or on other paved roads.) He wrote to various governmental authorities about a number of gun-related issues. (Guns, properly used by properly trained shooters, offered a source of pleasure and were useful in securing food.)

Jon also communicated with several Charolais breeders in search of an outstanding herd sire to stand in stud at Moccasin Lake Ranch. He felt that buying the right bull was the key to building a purebred herd with the genetic features most prized by potential buyers. Jon became increasingly convinced that owning a first-class purebred Charolais herd and selling yearling bulls would increase ranch income to that elusive break-even point.

He had read about the large sums being paid for champion Charolais bulls, especially at auctions in the Midwest. The top bulls sold for $100,000 and up, while bulls of less stature often brought $5,000 to $10,000. Titcomb felt that Moccasin Lake Ranch could develop a Charolais herd with credentials to compete in this latter arena. He concluded that with annual sales of fifteen to twenty young Charolais bulls, the ranch could achieve economic self-sufficiency.

Jon sent his son, Nim, and Roy Simons to Missouri in 1968 to look over the bulls being offered by the best Charolais breeders. Roy had by then succeeded Hod McKay as ranch foreman. Nim and Roy flew to Kansas City and drove to the farming community of Chillicothe, Missouri. There they visited Litton Farms, home to some of the best-regarded Charolais bloodlines in the country. They purchased from Litton Farms a grandson of the great sire Sam and also two purebred heifers.

The prized animals were carefully loaded into a large horse trailer and driven to Moccasin Lake Ranch to join the existing Charolais herd. Show pens were built adjacent to Patterson Lake Road in front of ranch headquarters. Jon felt that because the Charolais breed was not yet well established in the Northwest, it would be important from a promotional standpoint to keep the purebred animals where they could be easily shown to the public.

Jon wanted to provide better accommodations at the ranch to encourage visits by Nim and his sisters, Gaye (named after her mother) and Greata (named after her aunt). He decided to expand House No. 3 so it could accommodate the whole crowd of children and grandchildren. He added to the house, modernized the kitchen, and built a carport. The attic was set up in dormitory style to accept ten army cots outfitted with sleeping bags. This little turn-of-the-century cabin could now sleep sixteen people.

The renovations were completed in time to receive family members over the Christmas holidays of 1968. Upon their arrival, the family immediately got a stark introduction to one of the coldest winters in Methow Valley history. During the holidays, temperatures dropped to minus 42 degrees Fahrenheit and stayed near that extreme for several days. The winter of 1890–91 may have been colder, but no records remain to substantiate this. The winter of 1947–48 may have had more snow, but temperatures were less severe. Everyone's admiration for the early

Methow settlers was heightened when they reflected that pioneers survived the valley's winters without such common modern amenities as indoor plumbing, water heaters, insulation, and electric heat.

The bitter cold made year-end cheer difficult for all. Pipes under the house froze, and electric arc welding equipment was used in an attempt to thaw them. The wood fireplace was stoked around the clock to ward off the chill. The grandchildren slept on mattresses in the living room near the fireplace rather than in the unheated attic.

The severe cold hampered ranch operations and endangered the cattle. The tractors that hauled feed to the cattle in the frozen fields wouldn't start. The diesel fuel had turned into a jellylike substance that was virtually impossible to ignite. The crew did manage to start a gas-powered Willys Jeep. This was immediately pressed into service and used to tow the diesel tractors into the heated shop so they too could be started. The wood-burning stove so carefully positioned by Jon several years earlier was never more needed or appreciated. Once the diesel tractors finally came to life and warmed up, they were kept inside when not hauling feed to the livestock. The animals were fed greatly enhanced rations to help ward off the extreme temperatures.

Cattlemen were not the only ones ravaged by the winter storm. Orchards were devastated and to this day have not been replanted north of the town of Methow. That winter, which is still discussed in the valley, provided the kind of adventure that people want to go through only once.

It was Jon Titcomb's last Christmas at the ranch. He died eight months later, on August 21, 1969, in the Methow Valley town of Twisp. He was sixty-seven. His body succumbed to rheumatoid arthritis and the effects of years of the cortisone injections received for this condition. Jon's life ended much as he would have wished—quietly, on the ranch where he had made a difference, with a few family members and friends in attendance.

As a boy, Jon had experienced rural life in New England. There he had learned about the mysteries of the wilderness, gaining knowledge that he brought to his corporate career and to his time at Moccasin Lake Ranch. The ranch experience had provided both satisfaction and disappointment during eight eventful years. Through it all, Jon's overarching objective was to pass along to his heirs a self-sustaining asset that would provide to them, and to their heirs, peace, harmony, and beauty. There is no doubt that this latter objective was achieved. The wisdom of Jon Titcomb's original decision to buy Moccasin Lake Ranch is borne out daily by the vibrant activity at the ranch more than three decades after his death. As Jon had hoped, his grandchildren, and their children, come to play, and to rest, taking a recess from everyday cares.

Jon Titcomb with grandchildren Lisa (on left) and Paul. Moccasin Lake Ranch collection

The North Cascades Highway opened the Methow Valley to tourism when it was completed in 1972. Shafer Museum Collection

CHAPTER 13

The Highway That Changed It All

Citizen participation in community affairs reached new heights in the 1970s with lawsuits, the North Cascades Highway, and a new identity for the town of Winthrop. The last vestige of family farming gave way to tourists and part-time residents seeking escape from urban centers.

At the time of Jon Titcomb's passing, the winds of change were blowing through the Methow Valley. Family farms had been waning since the end of World War II, and now offered little more than a lifestyle and a part-time job. By 1970, larger spreads in the Methow were starting to be subdivided by real estate developers. The North Cascades cross-state highway connecting the Methow to the Puget Sound basin was becoming a reality. The only significant non-farm employment in the valley was at the Wagner sawmill in Twisp.

The North Cascades Highway seemed to be the key to the future of the Methow. It was to be a comfortable, but spectacular, two-lane band of asphalt through the alpine region grandly designated as the Pasayten Wilderness. The highway was completed in 1972 at a cost of $33 million. Instead of mining trucks and hay-hauling eighteen-wheelers traversing this most scenic of all Washington State mountain roads, the highway soon became popular with Puget Sound-area tourists who thrilled to its expansive alpine vistas. Although the motivation to take the northern route across the Cascade Mountains was initially fueled by its spectacular scenery, the 25 percent reduction in driving time from Puget Sound to the Methow Valley was also an important factor in generating traffic.

Completion of the highway was a boon to the Winthrop economy. The little town in the upper Methow was no longer just a way station near the far end of a country road. It now became the first town that tourists encountered after crossing the mountains. They would need gasoline and food, but merchants wondered if they could also be enticed to buy T-shirts, film, and trinkets.

In the period before the highway's opening, Kathryn (Kay) Wagner proposed that Winthrop adopt a unified Western theme along its four-block business district. She was the widow of Otto Wagner, the lumberman and sawmill operator, who had been tragically murdered a few years earlier. Kay Wagner calculated that the Western theme would boost jobs in Winthrop and help promote business. She envisioned a main street with buildings renovated and painted to look like those of early Western towns. Boardwalks would be built to enhance the visual impression. Western parades would further reinforce the image.

Wagner persuaded Winthrop's merchants to finance this renovation. She announced that if there was unanimous participation by the merchants, she would underwrite any funding shortfall. Every merchant pitched in financially. Bob Jorgenson, who had been instrumental in designing a Tyrolean village theme for the town of Leavenworth, Washington, was brought in to oversee architectural details. Artist Chet Endrizzi was hired to apply the Western look to the buildings' facades. When the North Cascades Highway opened in 1972, Winthrop was ready with its new look.

Tourists soon took to the eateries, shops, and saloons that dotted River Road, Winthrop's main street. What visitors saw when they entered town was a scene from the Old West — except that ice cream, sodas, and an assortment of knickknacks were within easy reach. Wagner's idea of decorating the town proved to be a stroke of genius. Visitors were entertained while patronizing the Winthrop shops before proceeding on to Sun Mountain Lodge or any of several other establishments providing overnight accommodations. The creation of a central theme helped save the town from the decay and obsolescence that has taken its toll on several other early Methow towns.

Winthrop adopted a new image in 1972 as a town out of the Old West. Shown is the town's main street and the annual 49ers Parade in 2003. Photo by Dennis O'Callaghan

In anticipation of the North Cascades Highway's completion, Jack Barron had begun planning the Sun Mountain Lodge resort a few miles south of Winthrop. Jack, who had taken over operation of the Sunny M Ranch from his father, Joe, in the mid-1960s, was well acquainted with Methow affairs. He persuaded state and federal agencies to help with low-cost loans for the lodge project as a way to provide jobs and help the depressed local economy. The plan called for an all-season resort perched high on the mountain, with a breathtaking view of the entire upper Methow. Construction started in 1966 under the watchful eyes of Jack Barron and renowned architect Roland Terry. In 1968 the Sun Mountain Lodge opened for business as the Methow's foremost resort.

Given its isolation on a remote mountain in the little-known Methow Valley, it is a tribute to the stunning architecture and unparalleled views — as well as the resort's amenities and its enterprising management — that Sun Mountain Lodge survived. The North Cascades Highway had not yet been completed, and the trip from the Puget Sound area entailed the crossing of two mountain passes and more than five hours of hard driving. Yet Sun Mountain successfully struggled through those early years on a split-season schedule. The summer season, May through October, was the major attraction. The brief winter season over the Christmas holidays catered to cross-country skiers.

Gradually the resort, which is just a fifteen-minute drive from Moccasin Lake Ranch, gained a following that allowed it to remain open for increasingly longer periods during winter. Skiing in the Methow was becoming more popular. Some enthusiasts used helicopters or snowmobiles to get into the rugged peaks to the north for downhill skiing in deep powder snow. Many others enjoyed the growing network of well-packed cross-country ski trails. A few years later, the resort began staying open year-round.

About the time Sun Mountain Lodge was opening, another recreational initiative was being planned some twenty-five miles to the north, not far from the village of Mazama. A group headed by Doug Devin began to study the feasibility of a ski resort with chairlifts for downhill skiing. The Methow Valley Winter Sports Council drew up a business plan and started gathering data on snow conditions. Early findings appeared promising.

The group took its ideas to the U.S. Forest Service, on whose land most of the skiing would take place. The agency seemed supportive. During the early 1970s, the sports council brought in the Aspen Ski Corporation to discuss creating a resort and developing the land around it. The corporation liked what it saw and, in 1974, took an option to purchase some twelve hundred acres of land at the base of the proposed ski hill.

Word quickly spread that the resort might ultimately service up to twelve thousand skiers and contain four thousand housing and lodging units. Land speculators entered the scene and started buying property. Not surprisingly, these activities worried some local residents. They began to realize that a project of this size and scope would change the valley's character forever. While some welcomed the year-round jobs that a large resort would bring, others felt that the environmental pollution caused by such large-scale tourism would more than offset any economic benefits.

Citizens and organizations in the Methow Valley and elsewhere took sides in an increasingly heated battle over the proposed Early Winters Ski Resort. The Aspen Ski Corporation, sensing endless confrontation, dropped out in 1977. It was immediately replaced by a local group headed by Doug Devin and his newly created Methow Recreation Inc. MRI was joined by Harry Hosey's firm, Hosey Engineering, in 1983. Development permits were applied for that year. The following year the Forest Service approved the project's environmental-impact statement. Meanwhile, opponents of the resort also gained allies: the Sierra Club, the Washington Environmental Council, and the Methow Valley Citizens Council all rallied around the opposition flag.

Against this background, the Early Winters controversy dragged on for another sixteen years. Emotions intensified and legal costs mounted with no resolution in sight. In 1985 a coalition of environmental organizations brought a lawsuit in federal court challenging the Forest Service permit that allowed development. The coalition won its case, but this decision was reversed the following year after an appeal in U.S. District Court. Then the Ninth U.S. Circuit Court of Appeals reversed the district court's decision in 1987 after concluding that certain aspects of the Forest Service's environmental-impact statement were flawed. Now the fight was in full sway and would not go away for another decade.

The Methow Valley saw other — lower-profile — changes in the years following completion of the North Cascades Highway. These included a school district consolidation that combined Twisp and Winthrop school programs at today's Liberty Bell facility. The Twisp school building soon was turned into a community center and library. Twisp also got its own police department after voting to discontinue Okanogan County's law-enforcement services. Meanwhile, Winthrop built a sewer system and treatment plant that was completed during the 1970s, and in 1980 the Methow Valley Ski Touring Association (MVSTA) was formed and began promoting cross-country skiing throughout the upper valley.

The Methow Valley economy evolved from ranching to real estate and recreation during the 1970s and 1980s. Photo by Don Portman, Sun Mountain Lodge collection

Grassy slope above the Wandling place. Photo by Scott Chambers

CHAPTER 14

Ranch at a Crossroads

Tough business decisions faced new management at Moccasin Lake Ranch. A fire, a land exchange, and a family decision to maintain ownership of the ranch helped formulate a business strategy for the future.

Our business strategy for Moccasin Lake Ranch during the 1970s was to simply carry on as an agricultural operation. We would continue to build both the Charolais and commercial cattle herds and to raise alfalfa for winter feed. An unspoken, perhaps subconscious, consideration was to minimize expenses, even if it meant gradually falling behind in the modernization of equipment and facilities. For the Pigott and Beatty families, Moccasin Lake Ranch represented a place to go for occasional visits. We drove to the Methow for long weekends, at the opening of hunting season, or during vacations as we had in the past.

By the mid-1970s, we recognized that maintaining the ranch's status quo was economically unsustainable. We came to the realization that the carrying capacity of the range exceeded the feed-producing capacity of our alfalfa fields. Fundamental changes were needed.

To help oversee the necessary changes, the family persuaded Nim Titcomb to move to Moccasin Lake Ranch in 1974 and assume management control. At first, Nim would work with Roy Simons, who had been the ranch's interim manager since Jon Titcomb's death. Roy was well respected, and had a thorough understanding of ranch workings. These two men worked well together. They both had years of hands-on agricultural experience. Roy was a lifelong farmer who had grown up and farmed on Montana ranches before moving to the Methow Valley. Nim started his career as a logging engineer with the Weyerhaeuser Company. After a few years of corporate life, Nim had joined with me in purchasing and operating both a cattle ranch at Creston and then an apple orchard near Chelan, both in Eastern Washington.

Nim and Roy considered business strategies that the ranch might adopt. Agricultural pursuits, in general, were difficult to justify economically. For example, we could buy hay on the open market to solve our alfalfa hay shortfall, but this was not cost effective. Another option was to find winter range in a milder climate. This too was uneconomical. Nim and Roy also considered, then rejected, the idea of establishing an apple orchard. The freeze of the winter of 1968 – 69 had destroyed virtually every orchard in the northern part of the valley. Orchardists felt it was too risky to plant replacement trees, particularly because the newer and more popular hybrids were judged to be less hardy than those that had been frozen out.

We made a collective family decision in 1974 to sell our cattle and lease out the rangeland to an independent operator, Howard Asmussen. Howard, a well-established cattleman, needed summer range for his existing herd, which he hoped to expand. Arrangements were soon completed for the range lease and to leave Asmussen's newly purchased heifers on our hillside pastures. Over the next year the remainder of our commercial herd was sold at auction. The Charolais cattle were sold individually as opportunities presented themselves.

The same year that he moved to Moccasin Lake Ranch, at the age of forty-two, Nim forsook bachelorhood and married Colleen Gilliam. Nim, Colleen, and the four youngest of Colleen's six children moved into ranch headquarters at House No. 1 and were soon enmeshed in ranch affairs. Typically, the season of the year influenced each day's work. Colleen reserved for herself responsibility for milking the two cows and caring for the chickens. Milking took place at twelve-hour intervals: shortly after 7 a.m. and 7 p.m. The chickens and pigs were fed after the morning milking. Then came the garden. Vegetables, berries, and fruit all required attention from May through September. Colleen was an expert at canning and preserving the season's bounty. To everyone's delight, she retains to this day her ability to produce top-notch pies and other delicacies.

During summer, Colleen's son Dale and her daughter Cheryl helped Nim and Roy with irrigation chores. Everyone pitched in to keep the various lawns mowed. To do this, the ranch's old lawn mower was hand-pushed a quarter-mile along the county road between houses. More than one driver was amused at seeing what at first glance appeared to be someone mowing the road's graveled shoulder.

In winter, schoolwork and related activities occupied the teenagers. Nim kept ranch roads clear of snow, fed the livestock, and ensured that the supply of firewood never ran out. This was important because the house's lack of insulation and the Methow's cold winters combined to make heating House No. 1 a considerable undertaking. Colleen remembers waking up one morning to find that a glass of water left overnight on the bedroom floor had frozen.

Gaye Titcomb passed away in Tacoma, Washington, in 1976. After the deaths of the elder Titcombs, the three Titcomb siblings — Nim, Greata Beatty, and my wife, Gaye — joined me as equal partners in the ownership of Moccasin Lake Ranch (I had purchased a 25 percent interest in the ranch from Jon in the mid-1960s).

As a hands-on manager, Nim Titcomb often oversaw Moccasin Lake Ranch from the seat of a tractor. Moccasin Lake Ranch collection

All but Nim were fully engaged in other business pursuits and were not involved in ranch operations. As a result, Nim was left to run the ranch on his own during the next two decades. He decided that the best course for Moccasin Lake Ranch was to focus on raising alfalfa for sale to dairy farmers on the west side of the Cascade Mountains. Additional revenues would be received from the range lease with Asmussen.

In order to increase alfalfa production, Nim upgraded and expanded the ranch's irrigation system. A portion of Thompson Creek was straightened and rerouted to the edge of the main fields. Nim then installed four wheel lines that improved productivity and allowed irrigation water to be distributed more evenly. Under Nim's supervision, these fields were also cleared of the old granary and potato cellar built by Frank Morse. Additional irrigation was installed around and above Moccasin Lake. The elevated ditch carrying irrigation water from Moccasin Lake was extended one-half mile eastward, providing water to still more alfalfa fields. With these improvements, the ranch was producing crops on more than two hundred acres of irrigated land, all gravity fed and without the need for wells or electric pumps. By 1980, the ranch's three summer cuttings were producing up to a thousand tons of alfalfa, its all-time record.

Nim was also working with aerial photographs and legal descriptions to more clearly establish ranch boundaries. This helped us to resolve issues arising from fences that were off the boundary lines either as a matter of convenience or as a

First the cattle herd was sold, then the rangeland was leased to Howard Asmussen in 1974. Moccasin Lake Ranch collection

result of inaccurate surveying. Nim's license as a professional surveyor and his background in civil engineering were important as we prepared to build miles of new range fences.

During this time Okanogan County approved the ranch's open-space land application. Open space was a designation that contained real-estate tax incentives. It was designed to encourage farm owners to keep their lands undeveloped and not subdivide them. Jon Titcomb would have approved of this land-use decision that preserved the ranch's open spaces.

In 1975, Moccasin Lake Ranch arranged a land exchange with Jack Barron that further enhanced its hay-producing capabilities, although it forfeited an opportunity to develop a woodlands subdivision. Barron proposed trading his irrigated alfalfa fields near the current site of Liberty Bell High School for ranch property near Patterson Lake. He wanted to subdivide this property for vacation homes, speculating that urbanites building weekend homes in the area would generate business for his Sun Mountain Resort.

The timing was right for both parties. When Barron approached us about a land trade, we had just sold our entire cattle herd. The hillside that Barron wanted to subdivide had been used exclusively for summer grazing but was now surplus to the ranch's needs. In contrast, the ranch's emphasis on increasing hay production made

Barron's alfalfa fields very attractive to us. These fields were easily accessible from ranch and county roads and conveniently located near other ranch fields.

After many meetings and negotiations, Jack Barron and Nim Titcomb agreed to an exchange in which Barron got the wooded hillside overlooking Patterson Lake, the house built by Eric Erickson in the 1920s, and a portion of the property originally settled by H. H. Williams at the top of the grade to Patterson Lake. Moccasin Lake Ranch received Barron's cultivated and irrigated fields on either side of Twin Lakes Road.

Each party got what made sense to it, given its business plan at the time. Barron anticipated the future growth of subdivisions and tourism in the valley. Moccasin Lake Ranch continued its focus on the production of alfalfa and on pursuing a course that would provide family members both recreation and business opportunities.

The next decade got off to a remarkable start in the Pacific Northwest with the eruption of Mount St. Helens on May 18, 1980. Favorable winds spared the Methow from the huge clouds of damaging volcanic ash that covered thousands of agricultural acres in Eastern Washington. As a result, the Methow's crops were sought after and most farmers benefited from exceptionally high prices that year. This bonanza lasted only one year, however, and most of the 1980s passed quietly at Moccasin Lake Ranch. There were few investments in equipment, facility, or crops. There were no additional land purchases and only infrequent visits to the ranch by the Pigott and Beatty families. Nim and Roy conducted the routine seasonal farm activities with the help of summer workers during harvest time.

Colleen was involved mostly with work in the vegetable garden, milking the dairy cows, and feeding the chickens. Occasionally she helped move cattle, herding them from one pasture to another when they needed fresh range. One unusual event that she clearly remembered involved moving Charolais bulls, truly huge animals, some weighing over three thousand pounds. A particular bull obstinately remained stationary despite Colleen's best efforts to move him. He simply refused to budge. As Colleen rode her horse, Paint, alongside the bull, she was horrified to watch the bull put his head under the belly of her horse and lift both her and the horse well off the ground. A moment later, the bull gently lowered his passengers to the ground and stared straight at Paint. It was then that Colleen decided she would be better off attending to another part of the herd.

One of the most enjoyable activities for the children during winter visits to the ranch was traversing its hillsides on the snowmobile. Over the Christmas and Presidents' Day holidays, there was always enough snow for cross-country snowmobiling. The children enthusiastically vied for time on the ranch's sole machine.

Moccasin Lake Ranch offered family and friends a variety of outdoor activities. Moccasin Lake Ranch collection

More than once a young person showed up on foot at the house, searching for help in extricating a snowmobile mired in deep snow. However, these occurrences did nothing to dampen the youngsters' enthusiasm.

Some ranch visits exposed us to unaccustomed activities, such as the fall ritual of beheading chickens and preparing them for the dinner table. After their heads were chopped off with a hatchet, the birds were dipped in scalding water and readied for plucking. Those who have personally experienced the work of plucking chickens will attest that it is both messy and painstaking. Not surprisingly, our children's urban classmates were generally unenthusiastic about participating in this aspect of ranch life.

In 1983 a startling event caused the Pigott and Beatty families to assess their feelings about Moccasin Lake Ranch and how it fit into family plans. On Christmas Eve, during extreme cold weather, fire burned the old T. D. Johnson house (later the first Frank Morse house) to the ground. The heaters being used in an effort to warm the house overtaxed its antiquated electrical system, causing the wiring to burst into flames. The Winthrop volunteer fire department responded to the call quickly, but once on-site the fire truck's equipment couldn't operate because its valves had frozen closed in the subzero temperatures. The house was close to Thompson Creek, but it too was frozen and couldn't provide water to fight the fire.

With Nim's stepdaughter, Cheryl Hunt, and her family safely outside the building, there was nothing to do but watch the flames consume the historic structure and all of the family's belongings. For the next few months, Cheryl and her family moved into House No. 1 with Nim and Colleen. That spring after the snows melted, a modern house trailer with a pop-out extension was put in place on the old cabin site, providing the Hunt family with a new home.

The fire of 1983 was the first on ranch property since the home built by H. H. Williams burned to the ground sixty years before. It's a wonder there were not more such incidents, given the tinder-dry, wood-framed cabins favored by settlers, the use of candles and kerosene lamps for lighting, and the wood-burning fireplaces and stoves.

After the fire, family members frequently discussed the ranch's long-term financial requirements. Sizable investments would be required to upgrade aging farm equipment, replace underground irrigation pipes, and renovate the houses. All these considerations were weighed against our assessment of the Methow as a place to spend time and its possibilities for agricultural development.

There was an open question within the family: if Jon Titcomb couldn't put Moccasin Lake Ranch on a profitable footing even though he diligently applied his business acumen and lifelong love of the outdoors to the task, what could we urbanites accomplish on a part-time basis? Jon's purchase of the adjoining Wandling Ranch had not redressed the economic shortfall. Neither had our subsequent liquidation of the cattle herd and pasture lease with Asmussen. It became apparent that one of our options was to follow the growing Methow Valley trend toward subdividing and selling land.

Our strategy was based on the fact that many residents of the Puget Sound area were now traveling to the Methow on the spectacular highway over the North Cascades. New weekend homes and ranchettes began to appear throughout the valley. With the closing of the former Wagner sawmill in Twisp in 1982, the Methow had no industry and a rapidly shrinking agricultural base. Tourism and recreation were becoming the valley's major economic drivers. Lifelong residents with low-cost land were increasingly tempted to sell it. Given these considerations, we thought it would be relatively easy to sell Moccasin Lake Ranch by simply placing it in the hands of a real estate agent. While no one was wildly enthusiastic about selling, family members took that first step in 1984 and had Moccasin Lake Ranch appraised to determine its market value. We then put the ranch as a whole on the market.

As time went by and no offers were received, we reversed our decision to sell. We found that most real estate activity in the valley consisted of small-acreage sales for weekend cabins or mobile homes. We agreed that breaking up the consolidated property that had been so painstakingly assembled over a century's time by

George Thompson and all of his successors, including Jon Titcomb, was a job that no one in the family had an appetite for. In the final analysis, Jon's vision served as a powerful argument against carving up the ranch into numerous small-lot subdivisions. For the second time in twenty years, Moccasin Lake Ranch was pulled back from the brink of being sold, and remained in the family.

The question of how to finance the replacement of farm equipment and facilities lingered. Some things had to be done immediately, such as installing new water lines to the new trailer house, the tack house, and the ranch shop. Other investments, such as replacing the haying machinery, could be spread over several years. Some of this equipment could be shared with others. For example, Moccasin Lake Ranch partnered with the Loomis Cattle Company (of which I am a part owner) to purchase a new hay baler. The new baler could produce 800-pound bales, a vast improvement over its predecessor's 100-pound bales.

To help finance some of these equipment purchases, Moccasin Lake Ranch participated in one land sale in March 1990. We sold 180 acres of upland range to John Blethan, who chose to replace the urban lifestyle of the metropolitan Puget Sound area with the rural atmosphere found in the Methow. Blethan and his family, who needed pasture for their horses, purchased Wandling Mountain acreage on our southeastern boundary. The sale provided Moccasin Lake Ranch with the capital to make necessary improvements to its infrastructure without significantly diminishing the carrying capacity of its summer range.

We also started looking for ways to expand and diversify the ranch's recreational activities beyond the fishing program with the state of Washington. In 1983 we agreed to begin hosting the Methow Valley Rodeo. Before that time, the rodeos over Memorial Day and Labor Day weekends had been held at the Sunny M Ranch. However, as rodeo popularity increased, the crowds outgrew the capacity of the Sunny M site. Sponsors looked for another location and found it at Moccasin Lake Ranch.

Ranch grounds across the county road from Twin Lakes were judged to be ideal. A large mound adjacent to an open field lent itself to spectator seating in tiers. The adjoining flat was large enough to house the rodeo arena, livestock pens, contestant warm-up areas, vendor kiosks, and plenty of parking. The arrangement resulted in a happy melding of traditional Methow cowboy activities and the valley's growing involvement with tourism and entertainment. After more than two decades at Moccasin Lake Ranch, the rodeo continues to return here twice each year, performing before ever-larger audiences.

Rodeo in the Methow was already a long-standing tradition when the site was moved to Moccasin Lake Ranch in 1983. Moccasin Lake Ranch collection

House No. 1 was renovated, new decks were added, and the garage was converted to a bunkroom for the growing number of Titcomb descendants. Moccasin Lake Ranch collection

Chapter 15

Paths to the Future

Diversification and subcontracting of some ranch activities allowed time for conservation work and modernization of ranch facilities. The Pigott and Beatty families spent more time in the valley and became active in both ranch and community affairs.

At the beginning of the 1990s, Nim Titcomb announced that he would like to sell his interest in Moccasin Lake Ranch. Both the Pigotts and Beattys agreed to buy Nim out on the condition that he continue to manage daily ranch operations. To further accommodate him, we built a new residence in 1991 for Nim and Colleen that overlooks Dibble Lake. This was the first new house built on the ranch in fifty years.

With Nim and Colleen relocated to their new home, the Pigott family moved out of House No. 3, which we had shared with the Beatty family for more than thirty years. We moved into House No. 1, while the Beatty family continued to use No. 3 for themselves and their guests.

Our move, although a simple relocation of less than a quarter-mile, was carried out with mixed emotions on the part of our children. They retained a sentimental attachment to the rustic old homestead cabin. They warmly remembered times spent there playing cards with their grandparents in front of the fire on long winter nights, playing in the alfalfa fields along Thompson Creek during hot summer afternoons, and sitting on the porch looking across the Methow Valley while listening to the adults discuss the day's events. They had even grown accustomed to the sound of bats flying out of the attic at night from their roosts over the army cots. Mice and chipmunks were daily visitors that were both a challenge and entertainment for the children during their visits from the city.

When Frank Morse built House No. 1, it was an imposing structure. The house had a full basement with a garage, dining room, kitchen, and a large cold-storage locker. Upstairs were three bedrooms and a living room, plus a kitchen and dining room. The masonry house was solidly built, with a relatively modern electrical system, but it was not well insulated. Nonetheless, we increasingly enjoyed our stays in this house and undertook various renovations, including a new deck and conversion of the garage to a downstairs bunkroom.

The size and number of fish produced by Moccasin Lake increased with intense management of its fishery. Photo by Scott Chambers

Now that the Beattys and Pigotts were spending more time at Moccasin Lake Ranch, our families continued to take advantage of its recreational opportunities. In winter, the rolling hills and dry snow made cross-country skiing a favorite pastime. In summer, when farm activities were at their peak, there was horseback riding for both ourselves and our children, who were fast approaching adulthood. And of course there was always fishing in Moccasin Lake.

At this time, the fishing program took on a new look. We decided to convert Moccasin Lake to fee-based fishing. For thirty years the state of Washington had stocked the lake with rainbow trout, and in return, the ranch had opened the lake to the public for fly fishing free of charge. But the state didn't have the resources to monitor the program. There was no limit to the number of fishermen who could be on the lake at any one time. There was no inspection of fish to determine their health, or of the fishermen to ensure they were not using artificial lures or bait. Further, there was no limit to the number of fish that could be killed.

We concluded that it would be beneficial to privatize the lake. Accordingly, in 1993 we exercised our option to cancel the agreement with the state and converted the lake to a fee fishing location the following year. We established a program designed to grow bigger fish than before. We also decided to limit the number of fishermen on the lake at any one time. Finally, there would be spring and fall fishing sessions of two months each. Professional guides would lend assistance as needed. They would also assure that only barbless flies were used and that all fish were returned to the lake in good condition. During our first commercial season, Jon Titcomb's grandson Ross Beatty worked as a guide at the lake and toured West Coast fly shops promoting the fishery. Ross, an experienced and avid fisherman, got the new program off to a great start.

To ensure that fishermen would have an enjoyable experience on the lake, the ranch stocked it with hatchery trout purchased from local suppliers. Fewer fish were planted, but they were bigger and of a different trout family. A Kamloops rainbow – steelhead crossbreed was selected because of its ability to tolerate the warmer surface temperatures of late summer. This hybrid also was popular because of its ability to grow to trophy proportions.

After a few seasons we re-engineered the water inlet to Moccasin Lake so that we would have the ability to control the lake's water level. Previously, lake water had been drawn down during the irrigation season as much as six feet. This did extensive damage to aquatic plants in the riparian zone and to the insects that feed on these plants. We started the renovation project by replacing Morse's fifty-year-old pipeline that carried water from Thompson Creek to Moccasin Lake. It had become plugged with debris in places and leaked in others. The new setup eliminated these problems and also allowed us to divert water directly to our irrigation system without any lake drawdown. In this way, lake water levels could be regulated and protection provided to the aquatic plants and insects.

These changes, along with new fencing that kept the cattle away from the lake's shoreline, contributed to dramatic improvements in the fishing program. Plant life and aquatic insects (fish food) suddenly appeared and prospered. Fishermen noticed that the fish were getting bigger and stronger. During this time, an artificial spawning channel was created under the direction of noted fish biologist Brian Chan. This channel permitted trout to spawn naturally in moving water, thus maintaining their health.

In the first decade of the new program, the average size of fish caught exceeded two pounds, although some much larger fish were caught each season. The increasing number of returning visitors each spring and fall is evidence that Moccasin Lake offers fishermen of all skill levels an enjoyable experience.

While we planned and carried out the various lake improvements, we were saddened to hear that longtime ranch employee Roy Simons had passed away at his California winter home on Thanksgiving Day in 1993. Roy's involvement with the ranch spanned a period of thirty years. He was one of Jon Titcomb's first hires in the early 1960s. For a time after Jon's death and before Nim's arrival from Chelan in 1974, Roy had managed Moccasin Lake Ranch, and he was active in ranch affairs until the very end.

Marge Simons continued to live on the ranch for two years after Roy's death. After Marge moved out, we realized it was not practical to modernize their ninety-year-old house near Dibble Lake. So we invited the Winthrop volunteer fire department to use the dwelling for fire-fighting practice. The invitation was enthusiastically accepted, and the house was burned to the ground as part of the exercise. Its demise left the uninhabited homestead cabin on top of Wandling Mountain as the only authentic pioneer structure remaining on ranch property.

Early in the decade, Moccasin Lake Ranch actively responded to the Methow's ongoing water shortage. The valley had endured a series of low-moisture winters. With the possibility of a drought similar to that of 1917–33, it was evident that we needed to conserve as much water as possible. Our range was being grazed during spring and summer under terms of a lease agreement with the Loomis Cattle Company. Water for these cattle as well as for irrigated pastures was critical. Nim began developing water holes wherever it appeared that water could be found. The ranch also started pumping water from Moccasin Lake and from wells to holding tanks on top of adjacent mountains. From these tanks, water flowed to strategically placed stock water troughs. These efforts greatly enhanced our range carrying capacity by giving us the ability to utilize our higher ranges, knowing that water would be available for the cattle throughout the summer regardless of annual precipitation in the valley.

Other water-conservation improvements followed later in the decade. More than a mile of open irrigation ditch ran from Moccasin Lake to an irrigation head box on Wandling Mountain. Seepage from the ditch was wasting water and eroding the hillside. To address these problems, we installed a buried pipeline to carry the irrigation water over the entire distance.

Perhaps we should not have been surprised to find that some of our conservation-related remedies had unexpected consequences. Although the buried pipeline eliminated water leakage, we belatedly discovered that deer, bears, birds, coyotes, and other wildlife no longer had access to water they had previously used from the open ditch. The leakage from the ditch also had supplied precious water to aspen groves that had grown up and prospered along its route. These were groves that provided shade from the summer sun for a variety of small game and also for our cattle.

To compensate, we installed a series of small holding basins along the pipeline. A network of perforated pipes was woven from the basins through the aspen groves to distribute the lifesaving water. This balancing of competing demands for water among trees, game, and crops no doubt will continue to play a part in setting ranch policy well into the future.

After twenty years of actively managing ranch affairs, Nim Titcomb told family members in 1994 that he wished to step back from the responsibilities of daily operations. His decision required the Pigotts and Beattys to make a strategic choice. Should we hire a foreman and continue to do the farming ourselves, or should we lease the cropland to an outside party as we had done with the rangeland? If we decided to continue our own farming, we would have to buy new equipment. Our windrowers, balers, and bale wagons were all more than ten years old, well into middle age by modern standards. If, on the other hand, we elected to lease out the irrigated fields, a reliable lessee would have to be found — one with proven agricultural skills and a willingness to sign a long-term agreement.

Ron VanderYacht turned out to be the person we were looking for. VanderYacht had come to the Methow Valley in 1971 from the little town of Lynden, which is located in Washington near the Canadian border north of Bellingham. He grew up there on a dairy farm and knew the dairy markets in the Puget Sound region. Ron planned to use the new North Cascades Highway for economical cross-state hay shipments while establishing a contract haying business to supplement his own cattle and farming operations in the Methow.

Ron began farming at Moccasin Lake Ranch in 1995. The first year, he concentrated on alfalfa. In subsequent years he introduced peas, barley, or wheat in selected fields, replacing the alfalfa crops previously grown there. Ron felt that crop rotation would improve the soil's condition. He also viewed crop rotation as a chemical-free method of reducing unwanted weeds. In the long run, Ron hopes to have Moccasin Lake Ranch's fields certified as organic. He is experimenting with a small stand of timothy. Hopefully, this effort will lead to new markets for ranch-grown products.

Since 1995, Moccasin Lake Ranch has not been directly engaged in either farming or ranching operations. With these operations contracted to independent operators, the family has become increasingly involved in Methow Valley activities outside the ranch. This is in keeping with the Methow's spirit and long-standing practice of volunteerism. Nim has been active with the Okanogan County Electric Cooperative, serving as a board member and as its president. He also has enjoyed long stints on the boards of the Wolf Creek Reclamation District and Winthrop's Red Barn. Perhaps most challenging has been his service as a member of the Okanogan County Water Conservancy Board. The board evaluates conservation programs and applications for water rights in light of the government's rigorous protection of endangered species. With the increased home-building in the valley and the attendant demand for more water, the board has an important but difficult task to fulfill.

A panoramic view that included Twin Lakes and peaks in the North Cascades was considered in siting our home in 1999. Photo by Molly Beck

After 1995 my wife, Gaye, and I began spending more time at the ranch than ever before. This led us to consider building a new home there. The remodeled main ranch house was comfortable but was becoming crowded. By 1995, our four children had given us six grandchildren, and we anticipated that the family might well get larger.

We spent almost two years evaluating possible building sites on the ranch's many benches and hills. We considered year-round site access, availability of water and power, sunlight at different seasons, and building costs. In the end, we selected a site on a hill overlooking the ranch in an area originally homesteaded by E. L. Boyd. Our north-facing outlook reveals the glories of the North Cascade peaks and their year-round snowfields.

The site for our new home is called The Ponds, in light of the two small bodies of water we created as part of the project. The ponds were excavated between two natural rises in the landscape that serve as the ponds' sides. Excavated dirt was used to fill in the open ends and serve as dams. The depressions were carefully outfitted with an impervious liner to prevent leakage and thus conserve water. We filled the ponds in late fall 1996, just before winter weather set in. The following spring we broke ground and began work on the buildings.

Construction of the buildings began in March 1997. It was a sizable project consisting of a main house, guesthouse, office and service building, plus lots of landscaping. We chose a western-style design that presents a low silhouette against north-facing Wandling Mountain. The buildings utilize rock and wood in conformity with the hues and shapes of the natural hillside. Aspen, pine, and spruce trees surround the compound.

When the work was completed in late 1999, we had an opportunity to thank everyone involved with the project. A party attended by the architect, designers, construction workers, landscapers, and their families was a highlight of the entire project. The celebration day dawned bright and clear, perfect conditions for a visit to our new home and its country setting. Before the day was through, a band played western tunes, to the delight of more than one hundred people who came to see the beautiful results of their labors.

One of Jon Titcomb's primary objectives at Moccasin Lake Ranch was to expose his urban-raised descendants to outdoor experiences. This was easily accomplished on a ranch, sometimes in spontaneous ways. For instance, there was the time when one of Jon's great-grandsons excitedly ran up with a cricket that he had just caught in the tall grass. However, the youngster objected to feeding the captured cricket to trout in the lake. He wanted to simply release it. When reminded that he fed crickets to his family's pet gecko back in the city, the child continued to object. He said the situation at home was quite different because the crickets there came from a store. It remained to be explained that the two crickets had a common beginning and would ultimately have a common end.

Such exposures to nature would have been of great amusement and interest to Jon, who delighted in watching his grandchildren participate in ranch recreation. However, it made little sense to him to maintain a stable of horses for exclusive use by his grandchildren. Accordingly, there was little recreational horseback riding during his decade on the ranch. The ranch horses at that time were spirited cow ponies, bred for working cattle under the reins of experienced cowboys. Gentler horses were considered a luxury that consumed pasture and winter feed while producing no meaningful work.

About the middle of the 1990s, we decided to build a more balanced stable of horses. We felt it was time to introduce Jon's great-grandchildren to horseback riding. There were now fifteen members of that fourth generation, most old enough to become equestrians, and therefore, we added horses suitable for beginning and intermediate riders.

Moccasin Lake Ranch became the home of Methow Valley Riding Unlimited, a non-profit program designed by Annie Budiselich that provides instruction for special riders of all ages. Annie Budiselich collection

One day we were introduced to a local nonprofit organization, Methow Valley Riding Unlimited (MVRU), and its director, Annie Budiselich (Annie B.). Annie and her husband had moved to the Methow Valley from the Puget Sound area. She brought with her a lifetime of riding, a love of horses, and specialized training for working with riders, especially children and those needing therapeutic assistance. Before her arrival in the valley, Annie had been involved with a popular riding program that included children of all ages. She hoped to establish a similar program in the Methow and was looking for a carefully constructed facility in which to teach.

We began a discussion about possible ranch involvement with MVRU. It would require construction of a special riding ring and other facilities: space for storing tack, a wheelchair ramp so riders could groom and mount their horses, a wheelchair-accessible bathroom, and space for meetings and training. The trade-off for providing these facilities was that Annie B. would be available to teach our grandchildren horsemanship.

Gaye and I elected to undertake the challenge. From 1998 to 2003, the various facilities were added piece by piece. The increasing popularity of the Methow Valley Riding Unlimited program has been demonstrated in many ways. A one-day riding jamboree at the ranch during the summer of 2002 attracted more than one hundred parents, friends, and children who rooted for their teams in various competitions. During that fun-filled day, all participants won prizes and everyone went home full of all the ice cream and cookies they could eat.

In recent years, the equestrian program at the ranch has been expanded to include MVRU-led therapeutic riding, private and group lessons, and for the more advanced riders, cross-country rides that incorporate numerous jumps over a specially designed course. Both ranch and MVRU horses are used as is appropriate. To further enhance these recent equestrian-focused activities, the ranch has purchased a four-horse trailer to access the nearby mountains, including the Pasayten Wilderness, with its spectacular vistas and miles of wooded trails. Our grandchildren are already eager to expand the scope of their riding to include visits to this delightful wilderness.

In concert with the new MVRU facilities, we rearranged and enhanced the ranch's main entrance during the late 1990s. Two new storage buildings were erected to house ranch equipment during the winter. The original ranch machine shop, designed by Jon Titcomb in 1961, was expanded and renovated. (Nostalgia compelled us to keep its wood-burning stove for heating, even though a more efficient infrared heating system was installed during the renovation.) Power lines in the area were buried, and a modern fuel center was installed. To complete the renovation, a new ranch entrance and sign were installed along Patterson Lake Road.

There have been other projects. In 2001 we built a new home for Curt Bovee, his wife, Lynn, and their two children. This house replaced the mobile home that had been a ranch fixture ever since the original Morse house burned to the ground in 1983. Curt has worked at Moccasin Lake Ranch for more than twenty years. His tenure is second in duration only to that of Roy Simons. Curt is a Methow Valley native who grew up on land near the ranch rodeo grounds. After school and military service, he worked summers at the ranch repairing fences. During winters, he drove logging trucks in the Methow. In 1990, Curt became a full-time Moccasin Lake Ranch employee and quickly became integral to its operations. In 2003, he became the ranch manager.

There seems to be no aspect of ranch life that is beyond Curt's knowledge or experience. In summer, he is busy managing the cattle and range or working on irrigation projects. In winter, he repairs equipment or busies himself removing snow. Curt recalls one summer's equestrian episode that he would just as soon have avoided. While he was riding one afternoon, his horse, Chillie, spooked for some unknown reason and began to buck. Almost instantly, the girth strap parted and the saddle (and Curt) flew off the horse's back. As luck would have it, the breast strap prevented the saddle from falling to the ground, and the saddle flopped over the startled horse's head. As it was being flung from side to side by the nearby berserk animal, Curt was left sprawled on the ground with a fractured vertebra. In due course, the animal regained its composure; Curt was able to summon help and was transported to the nearest medical facility. Fortunately, he had a complete recovery.

In recent years, Moccasin Lake Ranch has won county and statewide awards for conservation practices. This would not have happened without the contributions of Curt Bovee and Nim Titcomb, who guided the ranch through the final two decades of the last century with dedication and vision.

Nim Titcomb (on right) with ranch manager Curt Bovee. Moccasin Lake Ranch was recognized in 2002 by the Okanogan County Water Conservancy Board for its efforts to reduce water usage. Photo by *Methow Valley News*

Warm spring weather and a variety of wildflowers contributed to the growth of outdoor sports from mountain-tops to the Methow Valley floor. Photo by Don Portman, Sun Mountain Lodge collection

Chapter 16

A Delicate Balance

Moccasin Lake Ranch was in many ways a metaphor for Methow Valley life in the early twenty-first century. County planning authorities were confronted with land-use issues and the need for balance between the local economy and the robust environmental orientation of many residents.

Throughout its history, inhabitants of the Methow Valley have been witnesses to massive cultural change — a phenomenon that has left its mark on the valley at least four times since its beginning. Two centuries ago, Native Americans saw their ancient ways fade and then disappear altogether as explorers and pioneers appeared on the scene. A century ago, settlers arrived and successfully established their communities in the undeveloped wilderness. Fifty years later, World War II took many young people from the valley, hastening the decline of the family farm and starting the valley's transition from agriculture to recreation and tourism. More than three decades ago, the North Cascades Highway was opened and became a key factor in introducing a modern culture that embraces both part-time urbanites, with their cosmopolitan mores, and the rural traditions of the old Methow.

While absorbing the influx of urban newcomers during the 1980s and 1990s, the Methow had unfinished business to attend to. The dispute over the proposed Early Winters Ski Resort was entering its third decade. The issue had wound its way through the country's legal system en route to the U.S. Supreme Court. That court heard the case in 1989 without fully resolving it. The Supreme Court justices unanimously agreed that the Forest Service had the right to grant development permits on its land, but they also found its environmental-impact statement inadequate. The court ordered the Forest Service to expand its analysis of the impact that the proposed resort would have on the Methow's environment.

Another decade's time, millions of dollars in legal fees, passionate discourses in the media, and endless hearings and meetings all reflected the inflamed status that Early Winters had assumed in the public mind. The R. D. Merrill Company, a Seattle-based conglomerate that purchased the property at a courthouse auction in 1992, was well financed and enjoyed a reputation for being environmentally oriented. Many considered the company to be the Methow's best, last chance to develop an environmentally friendly resort. During its short ownership, Merrill constructed the attractive Freestone Inn, with its twelve rental units and a dining room overlooking a man-made lake. Merrill also renovated the cabins that Jack Wilson had built years earlier for his backcountry clients. These activities, however, represented only a small portion of the overall project planned by Merrill.

In the end, like previous owners of the Early Winters property, Merrill was never able to reach agreement with the environmental organizations that opposed large-scale development under any circumstances. The positions held by the two sides had become hardened beyond reconciliation. Finally, in December 1999, R. D. Merrill withdrew and the project was left unfinished.

Those who favor environmentally oriented priorities not only have managed to prevent the Early Winters Resort project from coming to the Methow, but also continue to oppose more recently proposed projects with job-producing potential. They have argued, among other points, that the valley's roads, schools, and utilities are already overcrowded or stretched to their limits. To this group, more seems to be less. Meanwhile, merchants and many service providers throughout the valley have encouraged tourism and are hopeful that population growth in the future will make a contribution toward a viable economy. Proponents of growth argue that, without an existing commercial or industrial base to provide year-round jobs in the Methow, working residents must either commute long distances or take second jobs to make ends meet. That the Methow has undergone a transition — with tourism and recreation replacing agriculture, mining, and logging — is no longer debated. But whether land-use and zoning policies should encourage or discourage population and economic growth is still an open and vital subject of discussion.

Continued investment in Sun Mountain Lodge contributed to its becoming the valley's largest employer. Photo by Don Portman, Sun Mountain Lodge collection

Should a four-season destination resort be developed? How about a small log mill to manufacture furniture? Is a pollution-free commercial call center acceptable? Does the Methow need a larger, uninterruptible electric power supply to support its ongoing growth? How about a high-tech incubator park? And most significantly, does the valley have enough water to permit new development while providing for the needs of endangered species?

Entrepreneurs who are willing to invest in start-up businesses in the Methow can point out that Jack Barron's Sun Mountain Lodge has been an important stimulant to the local economy for more than three decades with minimal adverse impact on the Methow's ecology. Under its new owners, the resort has expanded and upgraded lodge facilities in recent years, becoming the valley's largest employer. Years before Sun Mountain opened its doors, the Methow benefited from such early entrepreneurial investments as Colonel Hart's road, various mining ventures, Guy Waring's retail stores, and Otto Wagner's Twisp sawmill. Each such undertaking served as an important economic stimulus to the local economy by providing jobs and injecting much-needed currency into local businesses.

Perhaps most important of all is the debate over water and its allocation among various users. Water use is one of the tough issues facing the Methow's citizenry—especially in these times of drought. When the valley's first settlers arrived in the spring of 1887, they were thrilled to find tall green bunchgrass and see fresh-flowing streams running into a roiling Methow River. Today, springtime visitors are also impressed by the valley's bright green hillsides as well as the volumes of water in the Methow River and its tributaries.

However, the difference in water availability when the pioneers arrived and in the twenty-first century becomes quite apparent later in the summer. Nineteenth-century settlers arrived during wet years, when abundant snowpack, substantial rainfall, and minimal demand for water meant that the Methow River was relatively full even in the late summer or fall. But in the early years of the twenty-first century, winter snowpack and annual precipitation have been well below long-term averages. As a result, the upper Methow River has been virtually devoid of water by late summer, making it possible to walk across the river in several places. To the dismay of conservationists, these are the places in the river that represent prime spawning grounds for several endangered species.

Loss of spawning habitat caused by low water in the Methow has prompted water users, typically farmers and orchardists, to square off against government officials and conservationists. Implementing a sometimes controversial interpretation of the Environmental Protection Act, government officials have required many farms and ranches in the Methow Valley to reduce, or eliminate, irrigation in order to leave enough water in the Methow River for salmon, steelhead, and bull trout to spawn. As a result, crops have been reduced, orchards abandoned, and business decisions delayed. Some watersheds are closed to well drilling and residential development. The towns of Twisp and Winthrop are nearing the upper limits of their current water rights. Few would deny that the valley's recent water problems have taken a toll on both agricultural and commercial activities. On the other hand, many would argue that the water shortage itself has been exacerbated by inefficient irrigation practices and increased residential demand. As the Methow Valley's population grows and demand for water increases in future years, the issue of water use and its allocation will only become more formidable. Currently, the courts are reviewing the matter, evoking memories of the Early Winters Resort dispute.

Meanwhile, the passing years and improvements to the North Cascades Highway have brought a steady stream of new residents to the Methow. Many of these are seasonal residents who have moved into the unincorporated Upper Valley. Their arrival and home construction have not only placed increasing demands on the valley's water table in that area but also helped influence the local real estate market. When the North Cascades Highway opened in the early 1970s, Okanogan County appraised Methow Valley real estate at 14 percent of the county's total valuation. This was in line with the valley's approximate share of the county's population at that time. In 2000, the Methow's real property was appraised by the county at over 30 percent of the county total despite the Methow's 14 percent share (5,600 residents out of 39,500 in Okanogan County) of the population.

Ultimately, the issue that must be addressed by county planners is how to maintain the Methow's pristine environment at a time when Okanogan County's resources are stretched to their limit and year-round employment opportunities are in decline. If basic utilities become inadequate, classrooms overcrowded, or jobs limited to part-time self-employment, the valley will lose its special luster because county taxes will become insufficient to maintain roads, law enforcement, or other emergency services, and the quality of life here will surely deteriorate. The future well-being of Methow residents will depend on wise and balanced decisions by the valley's stewards.

The evolving economic and societal changes in the Methow have not bypassed Moccasin Lake Ranch. While we continue with ranching and agricultural activities, we have, as noted, incorporated conservation practices and diversification of ranch activities. Long gone is the water-wasting practice of flood-irrigating croplands. Reflecting the drought conditions of the early twenty-first century, the ranch replaced open irrigation ditches with buried pipelines, incorporated more efficient sprinkler nozzles, and lined ponds to reduce leakage. We have no doubt, however, that there is much additional work to be done in this area.

In the year 2000, we began to consider putting a large portion of our ranch land into a conservation easement. The objective was to keep a portion of our land in open space and permanently preclude its development. In 2001, we established a permanent no-development easement of approximately fourteen hundred acres. This was a step far beyond the earlier non-binding open-space designation that we had filed with the county for tax-abatement purposes. Working with the very professional staff at the Methow Conservancy, we created a plan that prohibits new structures on the

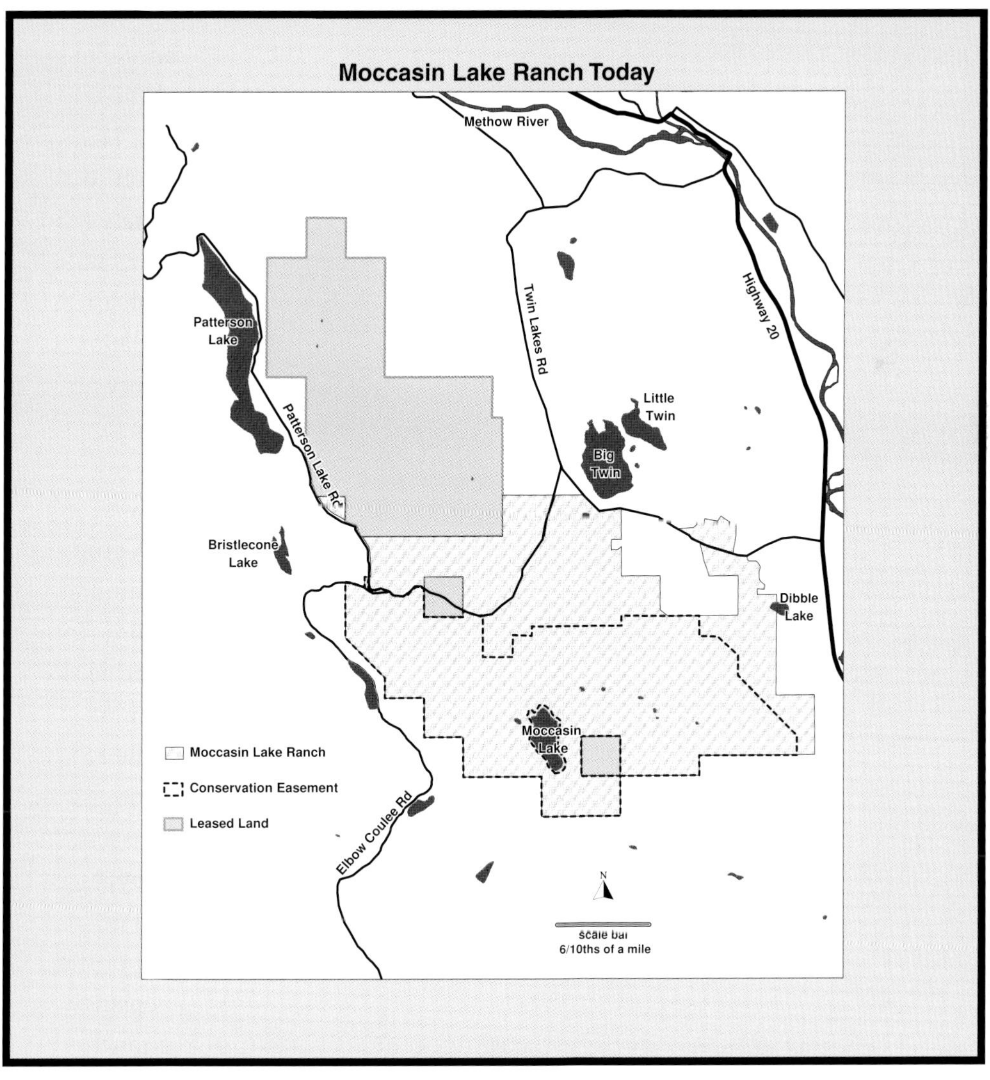

Cultural events became numerous with the growth of volunteerism in the valley.

designated land but permits recreational activities and cattle grazing. To preserve the Methow's look of undeveloped, open space, the land set aside for conservation includes all the ranch's ridges and elevated hillsides.

While there can be federal tax benefits from placing land in a conservation easement, these generally are secondary to the benefits stemming from maintaining or improving the quality of land stewardship. We felt that our ranching and farming operations could be maintained as in the past, while the conservation-easement lands would remain undeveloped and unspoiled in perpetuity. The Methow Conservancy has espoused these win-win concepts with considerable success in recent years. It deserves a great deal of the credit for steering us through the legal and accounting thickets that had to be traversed.

Since 2002, Kent Woodruff, a well-known valley wildlife biologist, has assisted Moccasin Lake Ranch with range management, noxious weed control, and wildlife habitat issues. While helping us in these aspects of ranch life, Kent has also been a great hit with our grandchildren. He has coached them on handling bats, calling owls and coyotes after dark, and tracking snowshoe rabbits during winter. These exposures to nature may seem routine to youngsters raised in rural areas, but for city dwellers, they represent insights into an exciting new world and a way to connect to a past that is fast fading from memory.

Jon Titcomb was quick to appreciate the Methow's open spaces and natural delights. As recounted earlier, he purchased Moccasin Lake Ranch so that his grandchildren, and their grandchildren, could experience simple rural pleasures. Jon Titcomb lived for only eight years after buying the ranch, but his care for the land has left an indelible imprint on his family. Though the activities at the ranch have changed over the years, Jon's interest in conservation has been vigorously maintained.

Just to the north of Moccasin Lake Ranch lies the Sullivan Cemetery, where many of the valley's pioneers rest. The names on the grave markers identify those who challenged the wilderness and won. They walked into the Methow, scratched out homesteads with minimal resources, and initiated a code of self-reliance and community effort that speaks to the valley yet today: live off the land; cure sickness; walk to destinations; stay warm in winter; protect your family; help your neighbor. Events of the day were discussed at Saturday night socials or literaries. Ladies' guilds made items needed throughout the community. Pioneers joined together to work on projects too big for any individual family. Volunteerism and community participation was ubiquitous in those early days because it was needed to survive in the harsh environment that confronted those intrepid people.

Geese in the fall sky above Moccasin Lake. Photo by Barry Provorse

Although the valley is now endowed with modern conveniences and is no longer wilderness, this spirit of the pioneers lives in the Methow in the form of energetic and widespread community involvement. More than seventy nonprofit organizations are at work in the valley. Methow volunteers support community activities such as alternative schools, live theater, ski trails, art galleries, concerts, rodeos, parades, and conservation efforts. Residents gather at community halls in Twisp and Winthrop throughout the year to address community needs. Volunteers operate and staff the libraries at Twisp and Winthrop as well as the historical museum and ice rink at Winthrop. As in the past, community involvement and volunteerism contribute immeasurably to the quality of life here. The valley's future as a desirable place depends on this continuing.

The Methow's attributes, a unique blend that includes clean air and water but extends to such high-tech innovations as Internet cafés and fiber optic data lines, give the valley vitality and variety. Though far removed from its roots of a little over a century ago, the Methow remains a special place for us all.

Source Notes

References and Suggested Reading

Davis, Robert Murray. *Owen Wister's West*. Albuquerque: University of New Mexico Press, 1987.

Devin, Doug. *Mazama: The Past 100 Years: Life and Events in the Upper Methow Valley and Early Winters*. Seattle: Peanut Butter Publishing, 1997.

Dibble, Dale. *Methow Valley Pioneers*. Naples, Florida: Privately published CD-ROM, 1998.

Ficken, Robert E. *Rufus Woods, the Columbia River, and the Building of Modern Washington*. Pullman: Washington State University Press, 1995.

Nugent, Walter. *Into the West: The Story of Its People*. New York: Vintage Books, 2001.

Portman, Sally. *The Smiling Country*. 2nd edition. Winthrop, Washington: Sun Mountain Resorts, Inc., 2002.

Waring, Guy. *My Pioneer Past*. Boston: Bruce Humphries, Inc., 1936.

Wilson, Bruce A. *Late Frontier: A History of Okanogan County, 1800 –1941*. Okanogan, Washington: Okanogan County Historical Society, 1990.

Wister, Owen. *The White Goat and His Country*. Winthrop, Washington: Shafer Museum, 1992.

General Periodical Articles:

Methow Valley News
Okanogan County Heritage
Okanogan Journal

Interviews:

Adams, John, Winthrop, Washington
Brewster, Perry and Dorothy, Twisp, Washington
Dibble, Dale and Olive Mae, Naples, Florida
Duffy, Barbara, Winthrop, Washington
Haase, Shirley, Twisp, Washington
Heckendorn, Henry, Bellevue, Washington
Huttell, Diana, Twisp, Washington
LaMotte, Vern, Carlton, Washington
Miller, Carl, Winthrop, Washington
Morse, Mary Jane, Wenatchee, Washington
Portman, Sally, Winthrop, Washington
Prewitt, Del, Twisp, Washington
Rader, Robert, Rio Rancho, New Mexico
Stokes, Jay, Twisp, Washington
Waller, Frankie, Winthrop, Washington
Webb, Richard, Ephrata, Washington
White, Donald, Winthrop, Washington

About the Author

Jim Pigott was born in Seattle, Washington, in 1936. He first became acquainted with the Methow Valley in 1961, when his father-in-law purchased a working cattle ranch located in the heart of the valley. A lifelong horseback rider and outdoorsman, Pigott has taken a keen interest in the Methow's history as well as in developments that have affected its current culture. Devoting increasing time to Moccasin Lake Ranch in recent years, he has played a significant role in aligning its ranching and farm operations with the Methow's evolving orientation toward tourism and recreation. The ranch currently makes its fishing and equestrian activities available to the public and also hosts semi-annual professional rodeos. In recent years, Moccasin Lake Ranch has won both county and statewide awards for its conservation practices.

When he is not in the Methow Valley, Jim Pigott lives in Seattle with his wife, Gaye, who shares his love of the valley and the ranch that her father purchased.

Pigott completed his undergraduate studies at Stanford University and then served two years of active duty in the U.S. Navy. He then attended Harvard Business School, before returning to the Northwest and pursuing a career in business. While at graduate school in 1963 he co-authored a book, *Integrated Circuits*, on the advent of silicon wafers, which were beginning to replace vacuum tubes throughout the world of electronics.

Jim Pigott, on a 2004 cross-country ride. Photo by Barry Provorse

Index

agriculture, 76, 99, 125
Alder Group Mining Company, 31
Alder Mine, 30-31, 32
alfalfa, 56
Arthur, President Chester, 25
ASARCO, 32
Asmussen, Howard, 132, *134*
Aspen Ski Corporation, 127-128
automobiles, 79-80
Azurite Mine, 31, 32

Bacus, George, First Lieutenant, 102
Bald Knob Trail, 44, 46, 47, 63, 67, 75
Ballard, Charles, 31-32
Ballard, Hazard, 32, 70
Bamber, Dan, 38, 40
Bamber Flats, 40
Bamber, Thad, 58
bank, first commercial, 67
Barnhart, Hily Ann and Addie, 86
Barron, Alex, 31, 32
Barron, Jack, 85, 86, 127, 134-135, 155
Barron, Joe, 102, 127
Barron, town of, 32, *33*
Bear Creek, 39, 40, 58, 60
Beatty, Greata, *13*, 110, 121, 132
Beatty, Mac, 115
Beatty, Ross, 143
Beaver Creek, 29, 36, 40, 41, 48, 58, 59, 65, 74, 85
Beaver Creek Cemetery, 41, 71
Beaver Creek School, 41, 59
Benson Creek, 36, 38, 44, 65
Berg, Charles, 117
Big Twin Lake, 116
Bishop, Minnie, 117
blacksmiths, 63, 69
Blanchard, C. H., 117
Blend, Dr., 102
Blethan, John, 138
Bly, J., family, 96
Boesel, Charles, 70
Bolinger, Walter A., 68, 73-74
Boulder Creek, 38
Bovee, Curt, 150, *151*
Bovee, Lynn, 150
Boyd, E. L., 84, 86, 90
Brant, Shorty, 35, 36
Brewer, Howard, 102
Brewster, 44, 47
Brewster Mountain Road, 44, 47
Brewster, Perry, 117
bridges, 51, *71*, 98, 99, 101
Bryan, Mott, 60
Budiselich, Annie, *148*, 149
Burgar, Amanda, 66
Burgars' Twisp Hotel, 63, 67
Burton, C. N., 76
Byrnes, James M., 29, 48, 65

Cap Wright Hill, 40
Carlton, 67-68, 75
cattle ranching, 77
cemeteries, 39, 41, 71, 158
Central Ferry, 44, 51
Chan, Brian, 143
Chancellor, 32, 64, 71
Chewack River, 38
Chewuch irrigation ditch, 77
Chiliwist summit, 38, 84
Chiliwist valley, 43
China Ditch, 27-28
Chinese immigrants, 27
Clark, James, 47
Cogswell, Tony, 39
Cole, Sam, 50
Columbia Basin, 24
Columbia Reservation, 24
Colville Indian Reservation, 24, 50
Commercial Bank of Conconully, 67
commercial enterprise, early, 63-65
community events, 59-61
Couche, Dr., 80
Countryman cabin, 58

Countryman family, 68
creameries, 77
Crockett, Henry, 117, *118-119*
crops, 55-58
Culbertson, T. H., 30
Curry, General A. P., 50
Curtis, C. C., 78
Curtis Sheep Slaughter, 78

dairy cattle, 77
Davis, E. R., 47
Davis, Jewett, 40
Devin, Doug, 127, 128
Dibble, Dale, 116
Dibble Lake, 85, 116, 117, 144
Dibble, Olive Mae, 48
Dillard, Carl, 68
drought, 78-79, 88, 89
dude ranches, 101-102

Early Winters Ski Resort, *103*, 127-128, 153-154
education, 58-59
Egbert, Arthur, 32
Elbow Canyon, 76, 90
Elbow Coulee, 86, 107-108
election, 1888 presidential, 60
electricity, 80-81, 99
Ellensburg, pioneers' trip from, 43
Endrizzi, Chet, 126
Erickson, Eric, 86, 90, 110, 135
Eureka Mine, 29, 31, 32

Fairview School, 59
farming, 53-58, 83, 108
Fender, A. C. (Linc), 67
Fender, Effie, 67
Fender, George, 76
Filer, Bertie Stone, 49, 60, 65
Filer, Flora, 51
Filer, Peter, 40, 47, 48, 67, 75
floods
    1894 flood, 51, 65, 66, *67*, *71*
    1948 flood, 101
flour mill, 76
Foghorn irrigation ditch, 41, 77
Ford, John and Rosa, 86
Forrester, Jack, 36, 65
Frazer Creek, 74
Frazer, Joe, 38, 40
Freestone Inn, 102, *103*, 154
freight service, 46, 47, 63, 87
French Creek, 59
Fries, U. E., 48, 69
Frisbee irrigation ditch, 41
Frisbee, Walter, 41, 47, 69
Fulton, Frank, 60
Fulton irrigation ditch, 77
Fulton, Lee, 40, 54, 58, 79
fur trading, 21-22

Garrett family, 86, 116, 120
Garrett, J. C., 85, 116
Garrett, Ras, 53, 116
German, Granderson, 41, 84, 86, 116
German, Will, 41, 86
Gilbert, 29-30
Gilliam, Cheryl, 132, 136-137
Gilliam, Colleen, 132, 135, 141
Gilliam, Dale, 132
Glover, H. C., 66
Glover, Jim, 36, 65
Gloversville. *See Twisp*
Goat Creek, 70
Gold Creek, 75, 76
gold prospecting, 27
Grant, President Ulysses S., 24
grazing permits, 78
Great Northern Railroad, 47, 69

Hart, Colonel W. Thomas, 31, 65, 155
Hartle, Elsie, 85
Hartle, John, 38, 39-40
Heckendorn, 70
Heckendorn, David (Gene), 70
Heckendorn, Louisia, 39, 69
Herstine, David, 51, 58, 85, 116
Herstine family, 59
Highway, North Cascades, 102, *124*, 125-127
Highway 20, 99-100
Homestead Act, 23
Hosey Engineering, 128
Hosey, Harry, 128
hotels, 39, 63-64, 68, 75
Houser, Pernina, 86
Hudson's Bay Company, 21

Huneke, Judge William A., 90
Hunt, Cheryl, 136-137
Huss, Ed, 41, 85
Huss, Harvey, 41, 84, 85, 116

Independence Day 1888, celebration, 60
Indians
  Chief Moses, 24-25
  Columbia Reservation, 24
  Colville Reservation, 24, 50
  hostilities with, 50
  Moses Reservation, 24-25
  treaties, 24
  U.S. Bureau of Indian Affairs, 24
irrigation, 55, 76-77, 90, 93, 95-96, 109, 113-114, 133, 143, 144, 155-157
Ives Landing. *See Pateros*
Ives, Lee, 63, 68

Johnson, Carl, 59
Johnson, Henry, 70
Johnson, Mary, 84, 88
Johnson, T. D., 83, 84, 87, 88-89, 96, 106, 120, 136
Johnson, Verna, 88, 90
Jones, Dick, 35, 36-37
Jorgenson, Bob, 126

Kopke, Gus, 107, 110

Lakeview School, 59
LaMotte, Vern, 61, 81
land consolidation, 83, 88-91
Laughton, Lieutenant Governor Charles, 50
Lawson, Dr., 80
Lewis, Buck and Boots, 29
Lewis, H. L., 76
Lewis, Harvey, 76
Libby, A. C., *34*
Libby Creek, *34*, 59, 75
Libby family, *34*, 68
Liberty Bell High School, 86, 116, 128, 134
Locust Inn, 68, 75
Long Jim, Chief, 35
Look, Charles, 69
Loomis Cattle Company, 138, 144

Mack Lloyd Park, 17, 69
Maddox, Wally, 40
Magee, William, 75
mail service, 48, 66, 69-70, 73
Malott, 78
Malott, L. C., 38
Malott, Mary, 38
Mammoth Mine, 32
Mansur, Charley, 36
Maxey, Cromwell, 47
Mazama, 70, 102, *103*
McAlister, Dan, 75
McCall, Jack, 93, 97
McCall, Mac, 97
McClure Mountain, 30
McClurken, Charlie, 70
McKay, Harley (Hod), 107, 120, 121
McKay, Mary, 107
McKinney, Buck, 102
McKinney, John, 63
McLean, Chauncey, 70
McLean, Dick, 75
meat, sale of surplus, 64
merchants, early, 64, 75-76, 155
Merrill, R. D., Company, 153-154
Metcalf cabin, 58
Methow Conservancy, 157-158
Methow Gold Company, 31
Methow Recreation Inc., 128
Methow River, 16, 41
Methow, town of, 29, 68, *75*
Methow Trading Company, *61*, 69, 70
Methow Valley
  agriculture, 76, 99, 125
  climate, 16
  economy, 100, 125, 137, 154-155
  electrification, 99
  employment, 100, 125
  flora and fauna, 17
  future of, 154-156
  geologic activity, 15
  glaciers, 16
  homesteaders, 37-41, 44-45, 48, 53
  inhabitants, early, 17-19
  land consolidation, 88-89
  population, 156
  recreational activities, 101-103, 127
  routes into, 43-44, 125
  smallpox, 19

soil, 17
tourism, 100-103, 125-129
transportation, 46-47, 73-74
20th century, early, 73
white men, first, 21
white settlement, 25
Methow Valley Citizens Council, 128
Methow Valley Riding Unlimited, *148*, 149
Methow Valley Rodeo, 138-139
Methow Valley Ski Touring Association, 128
Methow Valley Winter Sports Council, 127
mill, roller, 76
Miller, Carl, 89
Miller, Dick, 28-29, 31
mining, 27-33, 47, 65-66, 68, 70, 71, 100
Moccasin Lake, 83, 90, 95, 110-111, 113-114, 142-143, 144
Moccasin Lake Ranch, *14*, 83, 84, 86, 88-91
apple orchard, 95
business strategy in 1970s, 131-135
cattle, 109, 120, 121, 132
conservation easement, 157-158
farm equipment and facilities, 108, 109, 138, 149-150
farming, 95, 109, 133, 145
financial strategy in 1980s, 137-138
fires, 107, 136
fishing, 110-111, 114, 142-143
formation of, 83-90
horses, 148-149
House No. 1, 107, 137, *140*, 141
House No. 3, 110, 114, 121, 141
houses, 96, 97, 107, 110, 136, 141, 144, 147, 150
hunting, 115
irrigation, 90, 93, 95-96, 109, 113-114, 133, 143, 144, 156-157
land acquisition, 88, 89, 90, 115-117, 120
land exchange with Jack Barron, 134-135
land sale to John Blethan, 138
machine shop, 108
1970s, 131-135
1980s, 137-138
open-space land application, 134
potatoes, 95, 109
recreation, 114-115, 135-136, 148
rodeo, 138-139
sheep, 93-94
water pipeline, 95, 144
water rights, 84, 89-90
Morse, Bertha, 89, 91, 93, 96, 105
Morse, Bob, 89, 93, 95-97, 107
Morse, Donna, 97
Morse, Dorothy, 89
Morse, Frank, 86, 89-91, 93-97, 105-106, 107, 117, 120, 141
Morse, Frankie, 97
Morse, Marie, 97
Morse, Teddy, 91, 93, 95, 96, 97, 107
Moses, Chief, *20*, 25
Moses Reservation, 24-25
Mount St. Helens, 135
Mundy, Alva and Belva, 96

Nagley family, 117, 120
Nagley, Guy, 116
Nagley, Jefferson J., 116
Nagley, Louisa, 116
Nagley, Myrtle, 116
Nagley, Phoebe, 116
Narcisse, 36
newspapers, 74
Nickell, Alcena, 40-41
Nickell, Ben, 80
Nickell, Bud, 48
Nickell, George (Ed), 48, 75, 80
Nickell, Harvey, 37, 40, 48, 59
Nickell, Mary Ellen, 41
Nickell, Sally, 48
Nickell, Wash, 48
Nordyke, Eliza, heirs of, 90
North Cascade Mountain Range, 15
North Cascades Highway, 102-103, *124*, 125-127
North West Company, 21
Nosler, Charles, 68

Ogden, C. J., 68
Okanogan County Electric Cooperative, 145
Okanogan County Water Conservancy Board, 145, *151*
One Bar Ranch, 86, 116
Opera House, Twisp, 74-75
orchards, 55

Pacific Fur Company, 21
Paradise Hill Trail, 44, 46, 47, 63
Pasayten Wilderness, 125
Pateros, 37, 47, 64, 68-69
Patterson, Elsie Hartle, 40
Patterson Lake, 85, *87*, 89, 134, 135
Patterson Mountain, *14*, 15, 17, 86, 88, 90
Patterson, Sam, 40, 41, 84, 85, 86
Pearrygin, Ben, 38, 40
Pearrygin Lake, 40
Pennington, Charlie "Blackie," 63
Pennington, Frank, 63
Pennington, John, 63
Peterson, Nels, 63, 117
Pigott, Gaye, *13*, 121, 132
pipeline, water, 95
Polepick Mountain, 65
population, 53, 156
post office. *See mail service*
Prebylowitz, Frances, 51
Prewitt, Robert, 37, 40, 59
Prewitt, William, 40

Rader cabin, 58
Rader family, 59
Rader, George, 41, 49, 53, 68, 85, 86-87
Rader, Jack, 102
Rader, Pearl, 54, 64-65
Rader, Pleas, 41, 63, 68, 84, 86-87, 88, 90, 96
railways, 23, 47
Randall, Charlie, 63
Red Barn, Winthrop, 145
Red Shirt Mine, *26*, 29, 65, 66, 67
religious services, 60
Reynolds family, 68
Risley family, 68
Risley, Joshua, 75, 84
Risley, Lon, 75
roads, 73-74, *79*, 84, 99-100, 102-103, 155
Robinson Creek, 36
Robinson, Jim, 35-37, 69, 70, 75
Robinson, Tom, 35-37, 69, 70
Robinson, way station, 32, 64, 71
Rockview, 59, 63, 71, 76
rodeo. *See Methow Valley Rodeo*
Ross, Alexander, 21-22, 102
Rural Electrification Administration, 81, 99

sawmills, 63, *76*, 100, *101*, 125, 137, 155
schoolhouses, 59, *75*, 84
schools, 58-59
Shafer Historical Museum, *8*, 33
Sheehy, William, 117
sheep ranching, 78, 93-94
Sherman Silver Purchase Act, 65
Shulenburger, Clint, 102
Sierra Club, 128
Silver, town of, 29, 47, 48, 51, 65-66, 74
Simons, Marge, *108*, 110, 144
Simons, Roy, *108*, 110, 120, 121, 131, 135, 143
skiing, 101, 102, 127, 128
Skyline irrigation ditch, *77*
Slate Creek district, 29, 31, 65, 71
sluice boxes, 27
Smith, Al, 36, 37
social life, 59-61
Spaeth, Paul, 102
Squaw Creek, 29, 63, 68, 73, *75*
stage station, 67
stagecoaches, 47
State Game Department, 110
steamships, *42*, 47, 51, 68-69
Stevens, Jimmy, 96
Stewart, Jack, 70
Stone, Bertie, 40
Stone, John "Chickamun," 29, 36, 65
Stone, Manford, 40
Stone, Napoleon, 37
Stone, Nell, 60
Sullivan, Jimmy, 38, 39, 63, 69, 84
Sullivan, Louisia Heckendorn, 69, 84
Sullivan Pond, 39
Sun Mountain Lodge, 126-127, 134, *154*, 155
Sunday activities, 61
Sunny M Ranch, 101-102, 127, 138

telephone service, 74
Terry, Roland, 127
Tex, 107
Texas Creek, 44, 59, 75, 76
The Ponds, *146*, 147
Therriault, Leonard, 80
Therriault, Paul, 80

Thompson Creek, 38, 58, 86, 89, 90, 95, 96, 107, 136, 143
Thompson, David, 21
Thompson, Fred, 49, 84
Thompson, George, 38-39, 58, 60, 83, 84, 86, 88, 89, 96, 106
Thompson, Laura, 49, 54, 63, 84
Thurlow, Mason, 37, 40, 58, 60
Thurlow, Will, 40
Tingley, Minnie, 70
Titcomb, Colleen, 132, 135, 141
Titcomb, Gaye, 109, 110, 114, 121, 132
Titcomb, Jonathan R., 6, *7*, 11-12, 105-115, 117, 120-122, *123*, 147-148, 158
  cattle, 109, 111-113, 121
  hands-on management, 106
  irrigation, 113-114
  lawsuit against Frank Morse, 107, 111
  Moccasin Lake Ranch, purchase of, 105-106
  Wandling ranch, purchase of, 115-117, 120
Titcomb, Nim, *13*, 107, *114*, 115, 121, 131-137, 141, 145, *151*
tourism, 100-101, 125-129
transportation, 46-47, 73-74
Twin Lakes, 17, 29, 86, 138, *146*
Twisp, *62*, 63, *64*, 66-67, 74-75, 76, 80, 84, 100, 125, 128, 159

Upper Methow Valley Power and Light Company, 80
U.S. Forest Service, 127, 128

VanderYacht, Ron, 145
Vanderpool, George, 70
Ventura, 32, 64, 71
Ventzke, Albert, 70
Ventzke, Fred, 70
voting, 60

Wagner, Kathryn (Kay), 125-126
Wagner, Otto, 100, 125, 137, 155
Waller, Frankie, 93
Walter, Mathias, 87
Wandling, Edith, 116
Wandling, Gaylord, 116, 117, 120
Wandling Mountain, 86, 89, 90, 117, 138, 144, 147
Wandling ranch, 115-117, 120, *130*
Waring, Guy, *8*, 63, 69-70, 75, 76, 155
Washington Environmental Council, 128
Wassell, Bert, 105, 107
water cooperatives, 76
water rights, 84, 89-90
water shortages, 144, 155
Weeman Bridge, 76
Wehmeyer, Bill, 70
Wehmeyer, Fred, 70
West, Warren, 102
westward expansion, 22
wheat, 57
White Goat and His Country, The, 84
White, Joe, 40
Williams, Abbie, 58, 60, 83, 85
Williams cabin, 58
Williams, Clara, 86
Williams, Earnest, 88
Williams family, 59, 68
Williams, H. H., 84, 86, 135, 137
Williams, Newt, 47, 86
Williams, Robert E., 90
Williams, Susan, 86
Williams, Verna Johnson, 88, 90
Wilson, Jack, 102, *103*, 154
Winans, William Parkhurst, *23*
winter of 1889 – 90, 49, 54, 84
winter of 1947– 48, 121
winter of 1968 – 69, 95, 121-122, 132
Winthrop, 38, 39, 63, 69-70, 71, *72*, 75, 80, *98*, 125-126, 128, 145, 159
Winthrop, Theodore, 69
Wister, Owen, 84
Wolf Creek, 17, 18, 49, 63, 84, 89
Wolf Creek Reclamation District, 77, 89, 109, 116, 117, 145
Woodruff, Kent, 158
Woodward, Butler, 80-81
World War I, 80, 81
World War II, 99, 102
Wright, Cap, 38, 40

The Methow River in
fall. Photo by
Dennis O'Callaghan